Top Cover
for
North America

Defending America from the Soviet Threat

BY
TOM PETITMERMET

Contents

Foreword	xi
Chapter One Selected for Command	1
Chapter Two Galena History and Mission	9
Chapter Three Finally, in Command	15
Chapter Four Bringing Home the Mission	27
Chapter Five Getting Back to "Normal"	33
Chapter Six Arctic Freeze	45
Chapter Seven Picking Up the Pieces	57
Chapter Eight "Normal Operations"	67
Chapter Nine The Thaw Begins	95
Chapter Ten Disaster on the River	105
Chapter Eleven Personnel Problems on Steroids	119
Chapter Twelve More Personnel Problems	127
Chapter Thirteen Leadership Challenges	133
Chapter Fourteen More Leadership Challenges	147
Chapter Fifteen Time to Go	157
Chapter Sixteen Thoughts on Leadership	163

Foreword

Before I start telling the story of *Top Cover for North America: Defending America from the Soviet Threat*, I thought it would be wise to explain the meaning of that phrase. The motto of the Alaska Region of the North American Aerospace Defense Command (NORAD) and the Alaskan Air Command (AAC) was Top Cover for North America. This moto explains the role these two commands play in protecting North America from any potential Soviet aggression. The most direct route for the Soviets to attack North America would be over the polar route through Alaska and Canada. This mission of Top Cover was carried out by a very complex system of early warning radars, command and control facilities, airborne radar aircraft, and fighter aircraft on alert at remote operating sites in Alaska to intercept any Soviet attempt to enter U.S. or Canadian airspace. The story in this book is about my one-year assignment as the commander of one of these remote Forward Operating Locations (FOL), Galena Airport, that supported the Top Cover mission. The time frame for this story is October 1988 through October 1989 when the relationship between the U.S. and the Soviet Union was developing from overt Soviet aggression to a more peaceful détente or "thawing" of the Cold War arrangement between the Soviets and the U.S.

It is true in most organizations that the leader of those organizations spends 90 percent of their time dealing with 10 percent of their people. It certainly was true for my one year of leadership at Galena. I faced challenges that I had never seen during my prior assignments in the Air Force. However, the leadership tools that I learned during my one year of flying combat missions in Southeast Asia in 1971, flying fighters, testing air-to-air missiles, and standing air defense alert, certainly gave me some very good foundational tenants to lead my new squadron. The names of the 10 percent problem players have been omitted to protect the guilty. Not that the other 10 percent of my leadership challenges were not exciting. Most of the stories in this book revolve around some very unique situations. The stories run the gamut from the mundane to intense personnel issues to the downright exciting operation of a remote Alaska base to the air defense of the U.S. and Canada against the Soviet threat. I thought the best way to present these multifaceted situations was to relate them month by month.

Chapter One
SELECTED FOR COMMAND

One of the underlying goals of most, not all, U.S. Air Force officers is to become a squadron commander on the way to fulfilling many future high-level career goals. It has often been said that being a commander is the best job in the Air Force. I found that being a commander was certainly the best and yet one of the most challenging and rewarding assignments I had in the Air Force. Becoming a commander was one of my career goals during my twenty-six-year Air Force career. This book, *Top Cover for North America: Defending America from the Soviet Threat* is a summary of my memories when I served as a base commander at a remote Alaska base during 1988 and 1989. My career goal to become a squadron commander was formed by the incredible leadership I saw in the squadron commanders and other senior leaders I had the privilege of serving under both in war and peace early in my career. My memories are compiled based on a very detailed day-by-day journal I kept during this time. Additionally, multiple photos filled in a lot of the story. This journal/photo concept was a technique I used in my first book that I wrote in 2019, *Pretzel 06: Memories of a Forward Air Controller in Southeast Asia, 1970-1971*. The real impetus to write down my thoughts about my year in remote Alaska came from my thirty-eight-year-old daughter who asked me to "capture" the memories of this important assignment while I still remembered them. My daughter was only eight years old in 1989 and couldn't really understand why her dad had to be gone from the family for one whole year. The task of putting this book together was becoming a real challenge as I turned seventy-three this year (2020).

On my way to becoming a squadron commander I had a couple of very interesting career turns. While I was the Director of Readiness for the Alaskan Air Command (AAC) at Elmendorf Air Force Base in Alaska, I knew that the Air Force would soon be calling for me to complete a second unaccompanied remote tour somewhere in the world as it had been sixteen years since I returned from a one-year remote assignment to Southeast Asia. I expressed my desire to the AAC Personnel office to be considered for one the commander slots at one of Alaska's Forward Operating Locations. The two options were Galena Airport, Alaska or King Salmon Airport, Alaska. Since I had already pinned on my silver lieutenant colonel oak leaves, I knew that I was at least rank eligible to be on the command list. I still, however, had to work my way through the sometimes political "by name request" assignment process used by the Air Force and the Alaskan Air Command.

I was very familiar with the operations of both forward operating bases as my readiness director's job was to ensure that all of the approximately 10,000 Alaska Air Force and Air National Guard forces were combat ready and trained to do their mission. I had visited both forward operating bases many times during my readiness job and as the commander of an Alaska-based flying squadron and was convinced that I could do the job at either base.

So, in early January 1987 I got a call from the 21st Tactical Fighter Wing commander to make an appointment to interview for a squadron commander position in the wing. I put on my dress blue uniform and proceeded to the wing commander's office. There were three other lieutenant colonels, in their flight suits, waiting for their interview opportunity. I knew the competition would be tough as I also knew that all three other candidates were currently flying F-15s in the wing. I was the third in line to be interviewed and had a very positive interview with the wing commander. He asked all the right questions, and I thought I had all the right answers. Just as I was ready to complete the interview, he informed me that I had the job but would have to be flexible in scheduling a re-currency training class at Tyndall Air Force Base in Florida to become current again in the T-33 aircraft. I had flown the T-33 in 1970 at Cannon Air Force Base, New Mexico and again in 1972 and 1973 at Eglin Air Force Base, Florida. I was a bit confused by his comments and asked him if I would be allowed to fly while I was the commander of a remote site. He said, "Oh no, I have selected you as the commander of the 5021st Tactical Operations Squadron." I would be flying the T-33 aircraft from Elmendorf Air Force Base, Alaska.

The T-33 squadron, also known as the Aggressors, would be flying and testing the two F-15 squadrons in Alaska – utilizing tactics similar to what the

Soviet Air Force was flying and training with the F-15 pilots in Dissimilar Air Combat Training (DACT). Wow what a shock. I was getting a flying squadron commander's assignment but not a remote base commander's job. I was ecstatic as I never thought I would get to fly again in my Air Force career. I had flown nearly 2,800 hours (752 combat hours) in thirteen years to that point in my eighteen years in the Air Force. The T-33 commander's assignment lasted for over a year, and in the summer of 1988 the T-33 aircraft and mission was retired; my squadron flew all eighteen of the T-33 aircraft to the "Boneyard" at Davis Monthan Air Force Base in Tucson, Arizona. My time as the commander of the T-33 Aggressor squadron is another story still to be written.

Once the T-33 squadron was retired I was assigned as the Director of Inspections for the 21st Tactical Fighter Wing. On my second day on the job the wing commander asked me to stop by his office. I assumed he wanted to discuss my new job with him and his expectations for what he was looking for. To my utter surprise he informed me that he had selected me to be the next commander of the 5072nd Combat Support Squadron (CSS) at Galena Airport, Alaska, and I would be reporting in early October 1988. I had hit the jackpot, a second command and an opportunity to update my remote assignment status.

However, there was still another issue that needed to be dealt with. I was also selected in the spring of 1988 to attend the Air War College at Maxwell Air Force Base near Montgomery, Alabama and was to report there in August 1988. The commander of the Alaskan Air Command, a three-star general, told me while he was visiting my flying squadron that he thought I should attend Air War College. The incredible opportunity for a second command assignment was just too much and through some very professional whining, pleading, and begging, I convinced my boss, the wing commander, that I wanted to take my second command assignment at Galena. Fortunately for me he agreed, as I had already completed the Air War College program via the on base seminar program.

As I was getting ready for the big move to Galena, without my family, I had a number of things that needed to be accomplished. First and foremost, I needed to develop a plan for my family while I was gone for the full year. I immediately sent a request letter to the Elmendorf Base Commander to allow my family to remain in on-base housing at Elmendorf. We had moved into base housing in June of 1985 and the family was very comfortable with that arrangement. Fortunately, my request was approved, and my family remained in the same base housing unit while I was gone for the year. What a perfect arrangement as I had duties that would require me to return to Elmendorf from time to time for official business.

Next on the preparation schedule was attending the formal two-week long Base Commander's Course at Maxwell Air Force Base, Alabama. Since I had been stationed at Maxwell as an Air Command and Staff College faculty instructor from 1983 to 1985, I felt right at home at Maxwell except for the heat and humidity. The Base Commander's Course had a wide variety of students including many lieutenant colonels and full colonels on their way to their respective bases around the world. The course covered the entire spectrum of duties that a base commander may face. I was especially interested in the segment on the Uniform Code of Military Justice (UCMJ) since, as a commander, I would have UCMJ authority over the military members in my squadron. And believe me, it was wise to pay attention to that part of the training as I had to deal with a UCMJ action every single month I was at Galena. Other items covered in the course included the Air Force budget process, the Air Force promotion system, the Air Force awards and decorations system, transportation, maintenance, operations, and many more key things that a base commander would have to deal with. My head was rapidly filling up with many items that I had never had hands-on experience with.

Some of the foundational tenants I learned at the Base Commander's Course included the following from one of the many handouts (Air Force Instruction (AFI) 1-2, Commander's Responsibilities) I received at the course: Commanders are expected to display exemplary conduct as outlined in U.S. law, and all commanding officers and others in authority in the Air Force are required:

1. To show in themselves a good example
   of virtue, honor, and patriotism,
2. To be vigilant in inspecting the conduct of all persons
   who are placed under their command;
3. To guard against and suppress all dissolute and immoral
   practices, and to correct, according to the laws and regulations
   of the Air force, all persons who are guilty of them;
4. To take all necessary and proper measures, under the laws,
   regulations, and customs of the Air Force, to promote and
   safeguard the morale, the physical well-being, and the general
   welfare of the persons under their command or charge.
5. Commander's Duties and Responsibilities. Execute the
   Mission. Commanders hold the authority and responsibility
   to act and to lead their units to accomplish the mission.

6. Commanders must apply good risk management, accept risk and manage resources to adjust the timing, quality, and quantity of their support to meet the requirements of the assigned mission.
7. Commanders must ensure their unit is able to execute its primary mission at any time.
8. Within the scope of their authority, commanders must, at all times, maintain the ability to command and control their units against all relevant threats and hazards to assure mission success.
9. Lead People. Effectively leading people is the art of command. Commanders must maintain effective communication processes and ensure unit members are well disciplined, trained and developed. At all times, commanders must lead by personal example and pay judicious attention to the welfare and morale of their subordinates.
10. Further commanders will establish and maintain a healthy command climate which fosters good order and discipline, teamwork, cohesion and trust. A healthy climate ensures members are treated with dignity, respect, and inclusion, and does not tolerate harassment, assault, or unlawful discrimination of any kind.
11. Commanders must develop a two-way vertical and lateral communication system which is agile enough to respond to changes in the environment in a timely manner. In order to develop understanding, intent, and trust, commanders must transmit goals, priorities, values, and expectations, while encouraging feedback.
12. Commanders must cultivate a culture of compliance and accountability while promoting unit and mission pride. Command climate, customs and courtesies, uniform wear, physical fitness, and attention to detail are some indicators of the discipline of a unit.
13. Commanders must ensure their units are adequately trained. Unit training should take a building-block approach. Individuals must be proficient in career field specific skills before incorporating those skills into team and unit training. Unit training spanning the entire scope of the unit mission should include total force, joint, or partner-nation opportunities whenever possible. Training should replicate the distributed, chaotic and uncertain nature of expected operating environments.
14. Commanders will support the professional and personal development of subordinates. Professional development includes

formal mentoring, professional military education, academic opportunities, and other broadening opportunities. Personal development strengthens physical, mental, social and spiritual resiliency in an effort to build well-rounded Airmen.

15. Commanders have the unique authority and responsibility to engage in the lives of their subordinates, where appropriate, to improve quality of life, promote unit morale, and ensure all members are treated with dignity and respect.

16. Commanders are entrusted with resources to accomplish a stated mission. Those resources include: manpower, funds, equipment, facilities and environment, guidance, and Airmen's time. Commanders must consider risk in their stewardship of scarce resources to ensure effective and efficient mission accomplishment. As part of managing their resources, higher echelon commanders must ensure adequate resources are provided to subordinate commanders. Likewise, subordinate commanders must inform higher echelon commanders of resource shortfalls.

17. A commander's stewardship of personnel to meet evolving mission requirements is vital to mission success. Accurate reporting of manning levels, personnel rotations and readiness is vital when communicating with higher headquarters.

18. Commanders must base their budgetary decisions on mission requirements. Budgets must be credible, defensible, executable, and should contribute to cost-effective mission execution. Accountability and judicious management of funds must be command emphasis items aligned with command priorities. Make every effort to return excess funds to higher echelon commanders for reallocation.

19. Equipment and supplies must be properly accounted for, well maintained, and adequate for the assigned mission. Like manning and training levels, equipment status is a vital part of readiness reporting to higher headquarters.

20. Commanders must develop sustainable installations and implement appropriate asset management principles for built and natural assets. Regularly scheduled inspections, maintenance, and upgrades must be coordinated with appropriate agencies.

21. Unit members must have access to all command, technical, legal and procedural guidance necessary for mission

accomplishment. When necessary, commanders will publish guidance to document unit-specific processes and standards. Commander's intent is one vital piece of guidance commanders must provide to subordinates to ensure unity of effort.

22. Continuous process improvement is a hallmark of highly successful organizations. Wasteful, ineffective or unsafe ways of doing business cannot be tolerated. Commanders must foster a culture of innovation and challenge inefficiencies. A process for identifying and fixing deficiencies should be established and followed. Commanders must make data-driven decisions and manage risk while ensuring their unit's authorities, missions, plans and goals stay strategically aligned. A robust self-assessment program will identify the root cause of deficiencies and enable sharing of best practices with other organizations. Commanders are also expected to inspect their units and subordinates to ensure maximum effectiveness, efficiency, economy and discipline of the force are maintained.

I successfully completed the Base Commander's course and wondered how in the heck I could possibly follow those twenty-two tenants of command. I then tried to meet all the 21st Tactical Fighter Wing deputy commanders at Elmendorf for their advice on how to best support the wing and the assigned mission at Galena. I met with the deputy commanders for maintenance, operations, supply, resource management, and transportation, as well as the Alaska Air Command, command chaplain, and the judge advocate. I was more than ready to integrate all the wisdom I had learned from my previous nineteen years in the Air Force with the advice from these wing functional leaders into an efficient squadron at Galena.

Since I had been on the AAC staff for eighteen months and responsible for the war-time readiness of the command, I knew in great detail the mission that Galena played in the overall defense of the U.S. The Alaska Region of the North American Aerospace Defense Command's (NORAD) and AAC's primary mission was Top Cover for North America, or protecting the U.S. and Canada from any Soviet incursion into our joint airspace. Galena would have two F-15 fighters on immediate alert status twenty-four hours a day, seven days a week to respond to and intercept any Soviet aircraft attempting to enter the well-defined airspace.

I had flown many missions in and out of Galena during my time as the commander of the T-33 Squadron. I knew the challenge of flying in the arctic winter conditions in the interior of Alaska. I had an incident occur on one such mission, in January of 1988, as I taxied out for an early morning flight in -40°F weather. As I added power to taxi to the runway the crew chef frantically signaled me to stop. Both of my main tires had frozen to the ramp and when I tried to move, I taxied right out of the tires which were solidly frozen to the ground. I had moved forward about fifteen feet just on the tire rims. This incident was the result of a very hasty aircraft prefight inspection in the dark at -40°F.

We also had a contractor-operated, AN/FPS-117 Minimally Attended Radar at Galena. The role of Galena and the 5072nd Combat Support Squadron (CSS) was to provide all the support necessary to keep the F-15s in a position to launch for an intercept within a five-minute window and to ensure whatever support the contractor needed to efficiently operate the AN/FPS radar system.

In addition to the FPS-117 radar system, we also had two additional support units assigned to Galena. The first was Detachment 2 of the 1930th Communications Squadron that provided all airfield tower control and approach control radar support for both military and civilian flying activities in and out of Galena. Also assigned to Galena was Detachment 2 of the 616th Aerial Port Squadron that provided all loading and unloading of military supply support at Galena.

The entire support staff from the 5072nd CSS, and our two detachments, when fully manned, consisted of 320 military personnel and thirty-five civilian and contractor personnel. We had the complete cadre of support from operations, maintenance, civil engineering, supply, fire protection, personnel, morale and welfare, medical, the chaplain's office, housing, food services, airfield control, and aerial port operations. I will cover more of the specifics of our unique mission later in this book.

Chapter Two
GALENA HISTORY AND MISSION

The history of Galena Airport is long and storied. The Galena Forward Operating Location (FOL) is located on the northern bank of the Yukon River near the city of Galena, approximately forty-five miles east of Nulato, 270 miles west of Fairbanks, and 350 miles northwest of Anchorage. There are no roads into Galena, and all supply support either comes by barge down the Yukon river, when it is not frozen, or by air delivery. The city of Galena, Alaska was established in 1918 as a supply and trans-shipment point for lead ore (galena) from the mining camps south of the Yukon River. The location was on the site of a former Athabaskan fishing camp recorded in the 1880 census map as Natulaten. A school was established in the mid-1920s, and a post office opened in 1932. The population of Galena in 1940, the year the Civil Aeronautics Authority (CAA) began the first buildup in Alaska, was thirty people. Most residents were Athabaskan Indians who moved there from nearby villages on the Yukon River. The population of Galena during 1988-1989 was approximately 300 people.

In 1941, the federal government set aside 5,282 acres for the CAA to establish an air navigation facility as part of an overall civilian airport construction program in Alaska. The CAA selected Galena because of its central location in interior, western Alaska. Prior to World War II, the military had no presence in Galena. From August 1942 until September 1945, Galena supported the lend-lease transfer of American aircraft to Russia. Nearly 8,000 aircraft were flown through Galena on their way to the Soviet Air Force from 1942-1945. One remaining structure from the lend-lease program was a birchwood hangar, the largest building at Galena still remaining from the World War II. This hangar, in 1988-1989, housed our base operations section, aerial port section, aerospace ground equipment (AGE) section, some supply storage areas, and a very large open bay to park aircraft and vehicles in when necessary.

The official start of the Galena Airport began in 1942 as part of the U.S. effort to assist the Soviets in the war against Germany. The official name of the program was the Alaska to Siberia Lend-Lease-Program designed specifically for the transfer of military aircraft to the Soviet Air Force. Galena was a refueling stop for the transfer of these aircraft. Construction of support facilities for an Army garrison and support services was started in September 1942 but was delayed because of severe winter conditions until mid-1943.

After the war, Galena was returned to the control of the Civil Aeronautics Administration (CAA) in December 1945. It operated as a civil airport until early 1951 when the Alaskan Air Command (AAC) negotiated an agreement with the CAA for joint military and civil use. AAC planned to use Galena to base a squadron of fighter aircraft. The airport facilities were upgraded, but the squadron of fighters was never assigned to Galena, instead Galena became a forward operating base for fighters flown from Ladd Air Force Base in Fairbanks. Four Lockheed F-94s from the 449th Fighter Interceptor Squadron arrived at Galena on March 30, 1951. Additional improvements, including lengthening the runway from 6,500 feet to 7,250 feet and construction of a new combat alert cell and aircraft hangar that were completed in 1954. These upgrades were done to make the airfield ready for the Northrup F-89 fighters that replaced the F-94. Later, Convair F-102, Convair F-106, and McDonnell Douglas F-4 fighters would perform forward intercept duties at Galena. In October 1982, the arrival of the McDonnell Douglas F-15 meant very enhanced fighter intercept capabilities to counter the Soviet air threat.

The 21st Tactical Fighter Wing mission statement in 1988 summarized operations at Galena: "Galena Airport is the northern-most forward operating base in the Alaskan Air Command. It is the closest operating base to the Soviet Union. Its primary mission is to function effectively as an alert base. Two F-15s will be stationed on alert 24/7 at Galena. Their purpose is to intercept and observe unauthorized activities within U.S. airspace. The airspace that they will defend is known as the Air Defense Identification Zone (ADIZ)."

The ADIZ is an area surrounding much of North America, namely airspace surrounding the United States and Canada, in which the ready identification, location, and control of civil or military aircraft over land or water is required in the interest of national security. This ADIZ is jointly administered by the civilian air traffic control authorities, and the militaries of both nations, under the auspices of the North American Aerospace Defense Command (NORAD).

NORAD is a combined organization of the United States and Canada that provides aerospace warning, air sovereignty, and protection for North America. Headquarters for NORAD is located at Peterson Air Force Base near Colorado Springs, Colorado. The Cheyenne Mountain Complex in Colorado Springs housed the NORAD Command Center for a number of years until it was designated as the Alternate Command Center. The NORAD commander and deputy commander are, respectively, a United States four-star general or equivalent and a Canadian three-star general or equivalent.

The NORAD Command Center at Peterson Air Force Base, Colorado serves as the central collection and coordination facility for a worldwide system of sensors designed to provide the commander and political leadership of Canada and the U.S. with an accurate picture of any aerospace or maritime threat. NORAD has administratively divided the North American landmass into three regions:

    Alaska NORAD Region (ANR)

    Canadian NORAD Region (CANR)

    Continental U.S. NORAD Region (CONR)

The Alaskan NORAD Region maintains a continuous capability to detect, validate, and warn of any atmospheric threats in its area of operations from the Regional Operations Center Control Center (ROCC) at Elmendorf Air Force Base, Alaska. ANR also maintains the force readiness to conduct a continuum of aerospace control missions which includes daily air sovereignty in peacetime, contingency, and deterrence in time of tension and active air defense against manned and unmanned air breathing atmospheric vehicles in time of crisis. ANR is supported by both active duty forces provided by 11th Air Force, the Canadian Armed Forces (CAF), and reserve forces provided by the Alaska Air National Guard. Both 11th Air Force and the CAF provide active duty personnel to the ROCC to maintain continuous surveillance of Alaska airspace.

Canadian NORAD Region (CANR) Headquarters is at Canadian Forces Base (CFB) Winnipeg, Manitoba. It was established on April 22, 1983 and is responsible for providing surveillance and control of Canadian airspace. The Royal Canadian Air Force provides alert assets to NORAD. CANR is divided into two sectors, which are designated as the Canada East Sector and Canada West Sector. Both Sector Operations Control Centers (SOCCs) are co-located at CFB North Bay, Ontario. The routine operation of the SOCCs includes reporting track data, sensor status, and aircraft alert status to NORAD headquarters. Canadian air defense forces assigned to NORAD include the 409 Tactical Fighter Squadron at Cold Lake, Alberta and the 425 Tactical Fighter Squadron at Bagotville, Quebec. All Canadian squadrons fly the McDonnell Douglas CF-18 Hornet fighter aircraft. To monitor for drug trafficking, in cooperation with the Royal Canadian Mounted Police (RCMP) and the United States Drug Enforcement Agency (DEA), the Canadian NORAD Region monitors all air traffic approaching the coast of Canada. Any aircraft that has not filed a flight plan may be directed to land for inspection by RCMP and Canada Border Services Agency.

The Continental NORAD Region (CONR) is the component of NORAD that provides airspace surveillance and control and directs air sovereignty activities for the contiguous United States (CONUS). CONR is the NORAD designation of the United States Air Force First Air Force (1st AF), headquartered at Tyndall Air Force Base, Florida. 1 AF/CONR is comprised of Air National Guard fighter wings assigned an air defense mission to 1 AF/CONR on federal orders, made up primarily of citizen airmen, and its primary weapons systems are the McDonnell Douglas F-15 Eagle and General Dynamics F-16 Fighting Falcon aircraft. The organization plans, conducts, controls, coordinates, and ensures air sovereignty and provides for the unilateral defense of the United States. It is organized with a combined First Air Force command post at Tyndall Air Force Base, Florida and two Sector Operations Control Centers (SOCC) – one at Rome, New York, for the U.S. East ROCC (Eastern Air Defense Sector) and one at McChord Air Force Base, Washington for the U.S. West ROCC (Western Air Defense Sector). Both control centers are manned by active duty personnel to maintain continuous surveillance of CONUS airspace.

The FAA handles the requests of international aircraft and Transport Canada handles Canadian requests. Any aircraft flying in these zones without authorization may be identified as a threat and treated as an enemy aircraft, potentially leading to interception by fighter aircraft. Any aircraft entering an ADIZ is required to radio its planned course, destination, and any additional details about its trip through the ADIZ to a higher authority, typically an air traffic controller. The aircraft must also be equipped with a radar transponder.

The Galena runway was built right along the banks for the Yukon River. One unique aspect to maintaining the runways at Galena was the extensive dike system that was built around the entire perimeter of the runway and the military facilities on the airfield. Because of the very real potential for flooding during the ice breakup of the Yukon River, the dikes were designed to keep flood water out of the main operating locations. At the eastern end of the runway there was a simple system that kept the built-up dirt structure from eroding and causing damage to the runway. An innovative system of containers of liquid nitrogen were fitted on the eastern most edge of the runway. As the frozen soil on the end of the runway began to thaw, liquid nitrogen was automatically injected into the earth at -346°F, keeping the soil frozen and protecting against erosion. Later in this book I will relate an incident that happened in the spring of 1989 when the flooding from the river threatened the runway and the entire base infrastructure.

One other unique aspect of operations at Galena was the entire airport air traffic control (ATC) system was manned, maintained and operated by active duty Air Force personnel. This ATC cadre was comprised of approximately thirty members from a communications squadron detachment assigned to this control mission. The air traffic control detachment personnel manned the tower and approach control radar twenty-four hours a day, seven days a week.

In addition to the airfield and flying operations at Galena, the Air Force also had a AN/FPS-117 Minimally Attended Radar System in place on the base. The AN/FPS-117 Radar System is a 3-D (azimuth-range-height) phased array antenna radar. The system is a low-power, long-range (200-250 nautical miles), L-band pencil beam, solid-state transmitter and beacon interrogator search radar. The system design includes a redundant architecture with remote-controlled and monitored operations to minimize manning requirements.

The FPS-117 at Galena was part of the much wider North Warning System in place in Alaska and northern Canada. The AN/FPS-117 radars form an array stretching across North America from Alaska via Canada to Greenland. This North Warning System was designed to provide long-range detection and coverage for potential Soviet air threats and drug interdiction support and tactical command and control. Due to the extreme northern locations of these radars, the physics of radio wave propagation in the 1215-1400 MHz frequency range is even more critical for target detection requirements. This complete radar system was operated by six to eight contractor personnel with Galena providing support where necessary.

All of the command and control directions for our surveillances radar data and intercept activity was controlled by the Regional Operation Control Center (ROCC) at Elmendorf Air Force Base in Anchorage. The ROCC, in turn, coordinated with the NORAD Command Center housed in the Cheyenne Mountain Complex in Colorado Springs, Colorado. The F-15s on alert at Galena were the final "pointy end" of the air defense stick.

The Galena Air Force property had all the functions of a normal operating base. During the 1988-1989 time period there were approximately 300 to 320 active duty Air Force personnel and thirty to thirty-five civilian government employees assigned to the 5072nd Combat Support Squadron. In addition, there were between six and eight contractor personnel operating the FPS-117 radar system. All of the support functions that you would have on a normal Air Force base were in place at Galena.

Chapter Three
FINALLY, IN COMMAND
October 1988

Early on the morning of October 10 my incredible journey as the commander of Galena got underway. I was scheduled on a C-12 Flight at 0800 hours from Elmendorf with an en route stop at Eielson Air Force Base in Fairbanks. As we drove to the airfield, the roads on Elmendorf were very slick, and the van taking me to the flight line slid off the road into a ditch. Here I was on my way to assume command standing in the snow in -7°F weather helping the enlisted driver push the van out of the ditch and back on the road. The flight to Eielson was uneventful, and we had a very quick turnaround at Eielson. As we approached Galena, I noticed that portions of the Yukon River were already beginning to freeze. That freezing river would present me with some very challenging tasks as soon as I assumed command and again in the winter of 1988 and in the spring of 1989. After we arrived at Galena I was met by the current base commander and all six of the officers that would be working for me for a very short introduction lunch at the main dining hall. I knew that I would have my challenges at Galena because later in the day, when I was in the post office trying out the combination to my new mail box, the power to the entire base went out. Not a good start. As I was preparing to put my personal items in my Bachelor Officer's Quarter (BOQ) room, the heating system to my building went out as well. It was a very cold night (-10°F) as the heat did not come back on until later the next morning.

I had detailed discussions with the current Galena commander about some of the most important issues that I would be inheriting and expected to deal with. During a follow up lunch with all the officers the main dishwasher at the dining hall broke down and the entire base had to eat on paper plates for the next two days. Right after lunch I was walking through the base community center and noticed a hand-written note from the contract civilian barber at the base barber shop that she had quit that morning; we had no barber for the 320 military personnel on Galena. What a challenge for the change of command ceremony that would take place in two days. The troops were scrambling trying to get someone in the barracks to cut their hair.

Fortunately, there were a few barracks barbers that came through for us. I had a funny story later in the month after I took command. My hair was in need of a proper military haircut, and I asked around for the best barracks barber. A young enlisted supply sergeant was recommended to me by my first sergeant. I contacted the sergeant and asked if he could cut my hair. He agreed to meet me at the base barber shop. After he put the cape on me, he said "Colonel, I have never cut a white person's hair before." I told him to just do the best he could. The haircut turned out OK, but from then on I tried to get my haircut every time I was back at Elmendorf.

Bring My Airplanes Back. When I arrived at Galena the two F-15 alert aircraft had been deployed to Eielson Air Force Base in Fairbanks as the Galena runway and the combat alert cell (CAC) were both under major renovations. The two F-15 aircraft and fifty-four personnel from the 5072nd Combat Support Squadron had deployed to Eielson on April 30 to maintain the F-15 alert status for ANR. The squadron had sent thirty-eight maintenance personnel, twelve security police personnel, two barrier maintenance personnel, and two personnel from services (cooks). The deployment allowed the 5072nd to fulfill its mission of keeping the alert F-15s ready to scramble at a moment's notice while the runway and combat alert cell (CAC) were undergoing renovations. It would not be until November 30 that the F-15s and the support personnel would return to alert status at Galena. The road to getting the mission back to Galena has many more side stories that will be told in this book. Galena had the distinction of having the most intercepts of all of the Alaskan Air Command forward operating locations since we were the closest fighter base to the Soviet landmass. The deployment to Eielson was no different. Within two hours of the aircraft being on alert at Eielson, the F-15s scrambled for an intercept. In all, thirteen missions were flown from Eielson to intercept eighteen Soviet aircraft. This intercept story of Soviet aircraft would continue throughout my one year at Galena.

Leadership Opportunities. One of the first opportunities to show my leadership style was the day before the formal change of command. There was an announcement that today would be the day for annual flu shots, which were mandatory for everyone. So, first thing in the morning I went to the medical clinic and there were already a few people in line waiting for their flu shots. I pushed my way to the front of the line and said the new commander will lead by example. The head medical technician immediately came up with an idea. He would raffle off the opportunity to the highest bidder to give the new boss the flu shot. The bidding started at $10.00 for a shot in the arm and $20.00 to give the boss the flu shot in the butt. Needless to say, I was not too impressed with the idea of a shot in the butt but fortunately the only bidder, a senior supply sergeant, came up with the $10.00, and he gave me the shot in the left arm. This would not be the only time I had an opportunity to show my leadership style.

Change of Command. The formal change of command ceremony was scheduled for Friday, October 14 at 1000 hours in the base gymnasium at Galena. The 21st Fighter Wing commander officiated at the ceremony. All of the 5072nd CSS personnel who were not working or deployed to Eielson attended the ceremony. In addition, a number of Galena City Council members and the mayor of Galena also attended the ceremony.

The tradition of the change of command goes back to the time of the Roman Legions when the passing of the commander's baton occurred in front of the troops to identify the leader that would lead them into battle. During the eighteenth century flags were designed with distinct color arrangements unique to each unit and served as a rallying point and reminder of allegiance to their leader during battle. The Continental Army resumed the tradition for the U.S. military. The Air Force adopted most of the traditions founded by the U.S. Army after it became a separate service in 1947.

According to Air Force Instructions (AFI) 14.5.1, "The primary purpose of a change of command ceremony is to allow subordinates to witness the formality of command change from one officer to another. The ceremony should be official, formal, brief and conducted with great dignity." The AFI also stresses the importance of the flag/guidon exchange, stating, "The flag/ guidon exchange is exchanged during the change of command as a symbolic gesture providing a tangible view of passing of command. The sequence for the exchange begins with three participants standing abreast, from left to right as viewed from the audience, presiding officer, outgoing commander, and incoming commander. A fourth participant, the flag/guidon bearer, takes a position behind and between the presiding officer and outgoing commander. The flag/guidon bearer gives a subdued command, ('Officers, Center'); the presiding officer executes a left face while the two commanders execute a right face. The outgoing commander salutes the presiding officer, while stating, 'Sir, I relinquish command.' The presiding officer returns the salute. The outgoing commander takes the flag/guidon from the flag/guidon bearer, holding the flag and angling the top of flag towards the head of the flag/guidon bearer, and presents it to the presiding officer with presiding officer's hands on top. The outgoing commander takes one step to the right, two steps back, and one step to the left; simultaneously, the incoming commander takes two steps forward, taking the outgoing commander's place. The presiding officer presents the flag/guidon to the incoming commander who firmly receives the flag/guidon and passes it to the flag/guidon bearer. The incoming commander salutes the presiding officer, while stating, 'Sir, I assume command.' The presiding officer returns the salute. The flag/guidon bearer gives a subdued command, ('Officers, Post'). All three officers face forward.

The rendering of first salute/last salute is also an Air Force tradition and is appropriate during a change of command ceremony. The first salute/last salute is in keeping with the dignity and formality of the event. During the ceremony, airmen in formation, given the appropriate command by the formation commander, render a salute in place to show respect and appreciation to the outgoing commander through the commander's last salute from the unit. At the appropriate time after the passing of the guidon or the assumption of command, airmen in formation, given the appropriate command by the formation commander, render a salute in place to show respect and recognition of their new commander through the commander's first salute from the unit.

Following the formal ceremony, the entire squadron contingent, minus the former commander, moved to the Galena Consolidated Club for a reception and a meet and greet with all the city officials. I was finally in command and began an incredible learning experience that few military officers will ever have the honor of experiencing.

First Crisis. As I have mentioned earlier, Galena is only accessible by barge, when the Yukon River is not frozen between late October and late May, or by air year-round. The protocol for operating out of Galen on a year-round basis was to plan ahead for the next year's operation by ordering all the supplies and equipment the base may need for the next year. That pre-ordering process included basic supplies and equipment, food items, and, most importantly, diesel fuel to operate the large suite of generators that kept the power going at Galena and JP-4 jet fuel for the aircraft to operate and respond to Soviet air threats. We had six very large storage tanks that held all our reserve fuels. In the very first hour in command my head civil engineer (CE) officer came to my office and said the base would be approximately 700,000 gallons short of diesel fuel to operate our generators until the spring thaw because the Yukon River froze early. Our JP-4 fuel supply was exactly at the level we needed until the following spring barge delivery. The last of the diesel fuel deliveries was halted by the early freeze. One of the CE old timers recommended that we mix the diesel fuel with JP-4, allowing the generators to continue to operate, although less efficiently. However, one of the long-time power generator operators stated that using the JP-4 fuel mixture would significantly degrade the generators and burn them up. Obviously, we needed to come up with a solution to keep the base operating. My first call in command was to the resource manager (RM) at the 21st Tactical Fighter Wing to explain the real and inevitable problem of running out of diesel fuel for our generators. He stated that flying that much diesel fuel into Galena would be a massive dollar hit on the wing's budget. I told him not having the fuel for the generators would be an even bigger hit on the mission of the 21st TFW. My Galena CE experts told me we only had enough diesel fuel on hand to get us through mid-December.

The planning for the emergency fuel resupply started when it was confirmed that the barges could not make the trip to Galena in the fall. As a follow up to this problem, the wing eventually signed a contract with Markair Airlines to deliver the 700,000 gallons of diesel fuel. It took twenty C-130 flights to deliver the fuel under some very grueling weather conditions. The weather during the eight-day period between December 1 and December 9 varied from a low of -48°F to a high of -16°F. Through some very watchful eyes and highly professional work by all involved, we were able to transfer all of the necessary fuel to the storage tanks without spilling a single drop. The generators continued to operate throughout the year. However, later we faced another challenge when the weather at Galena reached a record -70°F.

Dining Hall Protocol. One leadership technique I learned from my time as the T-33 commander was to eat with the squadron airmen in the dining hall whenever I was there. So, other than during the Tuesday officer's staff luncheon, I sat with random members of the squadron to get a feel for how things were going. My first attempt was kind of funny when I sat with two very young supply airmen and asked them how their assignment was going. They were not very talkative, and I didn't hear so much as a peep out of them. I tried coaxing them to talk to me, but they had never seen their previous commander sit with the airmen and they didn't know how to act or what to say. I believe they were afraid to say anything. But I was persistent and finally after the third day when I sat with a group of Firedogs (fire department personnel) they opened up a little bit about how things were going. They did not like the music at the club or the food at the bowling alley, and the one thing they really did not like was the very small bowls for the ice-cream machine in the dining hall. Determined to fix that problem, I had them follow me to the ice cream machine where I grabbed a large water glass and filled it with soft serve ice cream. They said they weren't allowed to do that. I told them that starting that day they could use the glasses until we got larger bowls. Following lunch, I immediately stopped by the master sergeant in charge of the dining hall and told him to get larger servings bowls for the ice cream; I also told him to plan on making as much ice cream each day as his supplies allowed. Problem fixed!

Bird Poop Everywhere. Another lunch with the troops uncovered a real problem that should have been fixed long ago. I ate with some of the civilian employees that worked in the civil engineering section. They told me that their barracks windows were covered in bird shit, that they could not open their windows for fresh air, and that they thought it was a very unsanitary condition. As soon as lunch was finished, I drove the two civilian employees back to their barracks to look at their problem. Sure enough, it was very nasty situation. A certain type of swallow in Alaska makes its nest out of mud and had been doing so for many years on the barracks windows at Galena. The windows were covered in mud nests from the top to the bottom of each window, and bird shit was literally flowing down the windows. I asked them why this problem had not been fixed. They had complained about it for months but were told something about the birds being on an endangered species list. With all the command authority I thought I could have on my third day on the job I called the fire department on my hand-held radio and ordered the fire chief to bring the largest water spraying truck he had to the barracks. The fire truck was there in less than five minutes, and I ordered the responding firemen, the Firedogs, to spray all the windows in the barracks until all the mud nests and poop was removed. It was quite an undertaking, and years of mud and bird shit filled the street as the nests were washed away. I also called the head of civil engineering to meet me at the barracks and told him to keep up with the job of cleaning the windows for the squadron. As expected, I got a call the next day from the local Alaska Fish and Game Warden asking about my decision to clean all the mud nests from the windows. I explained, emphatically, that the health and well-being of my personnel took precedent over any bird, and that I would continue to keep the mud nests and bird shit off the barracks windows. Fortunately, he concurred with my decision.

During the remainder of the year I gleaned so much vital information from my very simple act of sitting down and eating with the airmen and civilian employees. I learned of some very vital mission areas where the airmen needed help, from supplies for the wood shop to transportation needs back at Elmendorf to problems that were not being addressed in many work areas. I also learned that seeing the commander come by the shops every now and then was great for morale.

Commander's Day Out. One other leadership technique I learned from a former commander was to get out of the office and see what the airmen were doing a on a day-to-day basis. I unofficially called it "leadership by wandering around." Whenever my schedule would allow it – I only missed about four weeks during the year – I would visit a single shop each week and spend the entire morning with the airmen. I didn't miss a single location. I did have some very interesting experiences with the airmen. While doing my half day with the fire department I got to drive one of the huge fire trucks, got to suit up in some firefighting equipment, complete with air tanks, and simulate a fire rescue with artificial smoke and all. It was quite the hard work for a forty-one-year-old lieutenant colonel. But the airmen were great and made allowance for the boss. I worked all morning in the supply warehouse sorting aircraft parts for inventory control. I was so surprised to find hundreds of parts for an aircraft, the F-106, that had not been stationed at Galena since the early eighties. My favorite location to visit was the dining hall where I would "help" the folks prepare the meals for the day. They always were concerned that "here comes the colonel, he will change the recipe again." I made it a point to never be off base during any holiday and I always worked the food service line handing out food to the airmen on the holidays. Many folks were very surprised to see me behind the serving line. I know how tough it was to be away from family during the holidays, and it was the least I could do for the squadron members and share the feeling of being away from family.

Visiting the Deployed Airmen. As mentioned earlier, the primary mission of the 5072nd Combat Support Squadron (intercepting the Soviet aircraft) was deployed to Eielson Air Force Base in Fairbanks when I arrived in October. They had been deployed since late April because the Galena runway was being completely refinished. It was time to visit the fifty-four troops supporting the alert mission at Eielson. I took a C-12 flight directly to Eielson on October 18. Prior to the flight to Eielson I got a call at 0300 hours in the morning that there had been a bomb threat at the combat alert cell (CAC) occupied by airmen at Eielson. The deployed operations officer at Eielson believed the threat came from one of the on-duty security policemen as the call went directly to the guard shack at the CAC. Oh, would this be a fun day!

I held a commander's call with the airmen at Eielson later in the afternoon. A commander's call is simply a meeting where the commander of a unit can talk to the personnel in his command to recognize outstanding performance and recent promotions, and equally important, get feedback from the personnel in the squadron. The commander's call was usually held in an auditorium, gymnasium, or base theater to accommodate maximum attendance and was considered a mandatory formation for all personnel not actively engaged in their mission.

I held my first commander's call in the aircraft alert hangar so that all fifty-four of the deployed personnel could attend. For the most part the morale of the airmen seemed OK. They were all very interested in when they could return to Galena. Following the commander's call, I met with the vice wing commander of the 343rd Fighter Wing and got a positive response to my question of how my airmen were performing their air defense mission on his base. He was very concerned about rumors that the re-deployment of our airmen was being delayed by some construction issues at Galena. I assured him that we were working all those issues to redeploy our deployed people back to Galena as soon as possible and that I was personally working every issue associated with the return to Galena.

Before finally departing for Eielson, I stopped by the Base Exchange store and bought two dozen doughnuts – a delicacy we normally didn't have access to – for our transient alert and base operations personnel at Galena.

Commander's Gram. Another leadership technique that I wanted to implement at Galena was the Commander's Gram Program. This program would give every person assigned to Galena, military and civilian employees, an opportunity to communicate directly with me on any issue that they had. I had a twelve-inch-by-twelve-inch box with a narrow slot at the top and a padlock on the cover installed in a prominent location in the base post office. I attached to the box a stack small forms with one question: "What do you want the commander to know or answer?" The line for the name was optional; however, without a name, I might not be able to get the answer directly to them. Everyone assigned to Galena visited the post office at least once a day to receive mail and care packages from home. The Commander's Gram Program box was installed on October 18. I was very curious what kind of response I would receive as I had mentioned the program at my first commander's call and mentioned that names were optional. Well, I didn't have to wait long.

The first submission was a two-page list from someone who identified each and every sexual relationship that was supposedly taking place on the base. Not only did I get names, I also got room numbers and if the people involved were married. Wow, what a hot topic to deal with. I called the chaplain to my office to discuss this issue, although we really didn't have an answer about what to do. We were both curious who the source of the information was and if that person was jealous of not being in a relationship. These relationships would certainly raise some leadership issues for me later in the year. For the time being, my leadership team was well aware of the potential sexual issues that could come up during a remote assignment.

The Commander's Gram Program seemed like a success because I usually received ten to twelve submissions a month covering a wide range of topics from promotion questions to requests for reassignments to specific mission support requests. I do believe that, in addition to my daily meals with the airmen and the weekly shop visits, the Commander's Gram Program helped me to answer many of the questions that were being asked by the airmen in the squadron and that they started to feel more comfortable talking to me during the meals.

Trouble Brewing. During my initial in-briefing from the previous Galena commander, he told me that he felt there was a hostile racial climate that was brewing at Galena and he had some concerns about some very basic issues between the black and white airmen in the squadron. This concern was also shared with me during my initial briefings with my six assigned officers and the top four enlisted members of the squadron. Specific individuals were mentioned who were believed to be causing the racial divide among the airmen. The potential for racial conflict at Galena was magnified when I got a call from the 21st TFW social actions office reiterating the previous commander's concern, which had been documented in his change of command after-action report. The social actions team wanted to come to Galena to conduct a unit survey on the racial climate on the base, which I encouraged them to do at their earliest opportunity. I wanted a baseline to see what direction and leadership skills I would need to exert to stop this "foolishness" immediately.

The signs were already visible when I heard a rumor that the new commander – me – had outlawed rhythm and blues music in the club, that whites would not let blacks use the dart board in the club, and, finally, a black female senior master sergeant had briefed all the young black female airmen not to date any white people while at Galena. I could see I would have my hands full at Galena as this was only the fourth day in command. This would certainly not be the end of this issue. There was a continued crisis brewing along racial lines that had the potential to destroy our entire squadron.

Red Cross Coordination. My immediate immersion into the very unpleasant task of coordinating with the Red Cross regarding deaths of family members back in the lower forty-eight states hit me on day three. Early one morning I got a call from the Red Cross informing me that the mother of a young airman had been killed in an auto accident in South Carolina. I had the unpleasant task of going to a dorm room and notifying the airman that her mother had been killed. Fortunately, my chaplain was with me for the notification, and he had all the right words to say. We also coordinated transportation for the airmen to fly back to South Carolina for the funeral.

Following that death notification, I had two more Red Cross notifications to do in the first two weeks of command. My deputy civil engineering officer was notified that his mother-in-law had passed away and he needed to get back to Florida for the funeral. Once again, the coordination with the Red Cross was superb, and we were able to get him on a plane the next day. Just one day later the Red Cross notified me that our chaplain's father had passed away on a hunting trip. We again scrambled to get the chaplain on the plane for the funeral. I was praying that we would have no more death notifications while our chaplain was gone.

No such luck. The day after our chaplain departed, I was notified that a young airmen's four-year-old son was very sick with meningitis. I started the process by going to the dorm and notifying the airman that his son was very sick and we were working on flights to get him home in time to help his wife care for his son. Just as the final coordination on the flight home was completed, I got a follow up call from the Red Cross that airman's young son had just died. That notification of the death was one of the toughest things I have ever had to do in my life.

One other very unique Red Cross notification occurred, again while my chaplain was gone and my first sergeant was on temporary duty at Elmendorf. I was notified that another young female airman's sister had been killed in a car accident in Florida. Since I had no chaplain or first sergeant available to assist in the notification, I attempted to contact the young airman's supervisor, a master sergeant. Unable to contact the supervisor I headed over to the young airman's dorm by myself. What a surprise when I knocked on the door…the airman's master sergeant supervisor answered the door in his underwear with a very sheepish and surprised look on his face when he saw who was knocking. Later that very same morning the master sergeant supervisor was no longer a supervisor as the Air Force, at the time, had some very strict rules against fraternization between superiors and subordinates.

Notifying a loved one about a family death is a very emotional and difficult task. The military doesn't pay lieutenant colonel commanders enough to do these terribly difficult tasks. I sure hoped my chaplain would return a soon as possible. I was not sure that I could handle another death notification. Unfortunately, there were many more deaths … and arrests and divorce notifications that I carried out over the next eleven months.

October Intercept Summary
    11 October : 2 Tu-95 Bear Hs
    12 October : 2 Tu-95 Bear Hs

Chapter Four
BRINGING THE MISSION HOME
November 1988

The rest of October was such a blur as I found myself chasing so many different programs/projects and dealing with additional personnel issues I thought I was in over my head as a base commander. First and foremost was getting the mission returned from Eielson Air Force Base. I thought that 5099th Civil Engineering squadron, the CE support squadron for all of Alaska, was slow rolling the upgrades to the combat alert cell (CAC), which was the last piece of the remodeling puzzle at Galena. The runway resurfacing and barrier upgrades were completed on time, and we were ready to bring the F15s back to Galena by the middle of November. But I had a number of additional challenges to deal with.

Combat Alert Cell (CAC) Upgrades. As part of the original remodeling plan the CAC needed extensive work from upgraded windows to new communication equipment and an enhanced alert alarm system to upgraded sleeping and eating rooms. The CAC was the central location where all mission activity starts. The CAC had sleeping quarters for the F-15 alert pilots, sleeping quarters for on alert maintenance crews, and a control cab for where our weapons and communication experts communicate with higher headquarters twenty-four hours a day seven days a week for intercept activities, a dining facility where the alert pilots and support crews ate, and dorm style rooms for sleeping. We also had a four-bay aircraft hangar that housed our alert fighters and some of our critical support equipment. The established alert launch criteria was that the alert crews would launch the alert aircraft within five minutes of a scramble order. The CAC had been in use since the late fifties with very few updates.

The 5099th Civil Engineering Squadron, headquartered at Elmendorf, was coordinating the rehab work and was really feeding me a bunch of baloney on the project. They were telling me the new windows for the CAC would take three months to acquire and then two weeks to install, that there was no money for new furniture in the CAC, and that they had no money to upgrade the esthetic look of the CAC. That gave me a real challenge.

I directed my squadron operations officer, who was deployed to Eielson with the aircraft, to go into the city of Fairbanks and see if he could find the exact windows we needed to upgrade the CAC. In less than two hours after my request he found a source in Fairbanks for the windows and said he could have them to Galena in two days. I made a command decision and told him to purchase them on the deployed credit card he had with him and to coordinate for delivery as soon as possible. The windows cost the squadron $3,200 dollars but were well worth the investment. The windows arrived two days later, and my own civil engineering experts installed the windows in a single day. Problem solved!

The next problem to solve was an upgrade of the furniture, both in the CAC working area and also in the sleeping and eating quarters, and to acquire some artwork to upgrade the entire look of the CAC. I asked my services officer to come with me as we scoured every dorm building on the base and found some beds, dressers and night stands for the sleeping quarters; they were not brand new, but most of it was in great condition. The bedside lamps were in OK condition, but we needed new lamp shades for each sleeping room. Again, I tasked one of my senior non-commissioned officers (NCO) who was back at Elmendorf to find some lamp shades in Anchorage for the CAC upgrade. She was successful, and we now had the "new" furniture for the sleeping rooms.

My next focus was new tables and chairs for the eating area of the CAC. I got together with my food services expert and found three table and six chairs that could be utilized in the CAC feeding area. The tables and chairs weren't brand new but were much better than the existing furniture. Finally, during the walk through of each building on base, I identified twelve pieces of artwork, mostly photos of the Alaska landscapes, and "procured" them for the CAC. I asked a couple of senior Non-Commissioned Officers (NCOs) to artistically place them throughout the CAC. The place looked great.

Still on the to do list was completion of the power upgrades and new communications equipment in the CAC. The power upgrade was behind schedule, and the communication upgrades were also lagging behind. One evening I was at a city council meeting and got an emergency call that the power upgrades in CAC had caused a massive power outage on the base. Great, not only was the power to the CAC upgrade screwed up but power to the entire base was also in jeopardy. I immediately called an emergency meeting in the birchwood hangar, because it had back up power, and got all the engineers from the 5099th squadron and my civil engineer experts together and told them to identify the problem and come up with a solution. If they needed additional help, I told them I would call back to Elmendorf for guidance and assistance if necessary. Fortunately, the engineers identified the problem – a short in the main control panel – and had the power upgrade project back on track. The power upgrade was completed and certified in time for the airplanes to return on schedule on November 30.

The communications issue was a minor problem but it still concerned me. The communications upgrade was being conducted by a civilian contractor associated with a large a company in Anchorage. They had a contract incentive to get the communication suite working or lose money on a fixed-price and fixed-date contract. We were able to get the communication upgrades working and tested before the airplanes came back from Eielson. I passed the completion of the communications test results back to the 21st TFW and told them we were ready to bring our mission back to Galena. The wing informed me that they still had to coordinate airlift support to bring the personnel and equipment back from Eielson.

The deployed airmen and all their support equipment returned to Galena over a three-day period (November 27-29). With all the airmen back in place we ran through multiple exercises at the CAC to ensure all was ready for the return of our alert mission. On the 28th we had two jets fly into Galena to test the replaced barrier system. While there were a few glitches with the barrier, the crew quickly made repairs and practiced their procedures for future use of the barrier. We were now ready to start our alert posture and our core mission on November 30. The two jets and the pilots were ready to support the NORAD Air Defense Alert Mission from Galena once again.

United Service Organization (USO) Randy Travis Show. The USO had announced in early October that Randy Travis and Patty Loveless would be doing a show at Galena on November 30. It was quite an honor for our small base to have such a famous person perform there. On November 28 Randy Travis' front man arrived at Galena to coordinate all the technical support for the show. We showed him the gymnasium where the show would take place, coordinated all the power and light support he would need for the show, and ensured that he was satisfied that we could support all the requirements for the show. He expressed to me that his only concern was if the base commander would support the show by doing the introduction. He was very embarrassed when I told him I was the commander, and I would do whatever the show needed me to do.

At a city council meeting on November 20 the city leaders asked me for one hundred Randy Travis tickets because they wanted to sell them to raise money for a local resident that had been injured in a snow machine accident. I told them in no uncertain terms that they could not sell the tickets because we had a limited number of seats (275) in the gymnasium and my airmen would have first say on attending the show. I did say that if there were any tickets left over from my airmen, I would offer them to the city but under no circumstances could they the sell them. The city leaders were not too pleased with my decision; but that was just tough…it was the way it would be. In the end, I was able to give the city fifteen tickets to the show.

On the morning of the show I greeted the C-12 aircraft that brought Randy Travis, Patty Loveless, and their staff to Galena. I took them to the dining hall for lunch, and they wanted to go through the food line just like the airmen. They also wanted to sit in the main part of the dining hall so the airmen could come by and get their autographs and say hello. They were all very friendly and really made a hit with the airmen.

After lunch they asked to go to the town of Galena for some shopping. I piled them all in my command vehicle, a 1984 Chevrolet Suburban, and off we went to see both Old Town Galena and the new city Galena. As we walked through the one and only market in Galena the locals were stunned to see Randy Travis shopping in their store. The people following him increased ten-fold as the day progressed, but Randy was very generous handing out autographed photos of himself. Patty Loveless did the same. I got them back to the quarters at around 1500 hours and left them alone to prepare for the evening show.

I stopped by the gymnasium about an hour before showtime and met with representatives from the Country and Western Music Channel that would be televising the show live later that evening. They told me where I needed to stand to introduce the show, what not to do on live television, and where the flood lights would be focused when I spoke. I certainly was a bit nervous, but it was no big deal. When it was time to introduce the show, I first made a very important announcement on live TV that as of 1519 hours Galena was once more an active fighter base as our two F-15s assumed alert status today. The airmen cheered with pride that they would once again be supporting our important Top Cover mission from Galena.

The only concern I had introducing the Randy Travis show was with my officers. Earlier in the day, they kept asking me if Patty Loveless was not in fact the porn star Linda Lovelace from the famous movie *Deep Throat*. I was so concerned that I would flub up by introducing Linda Lovelace rather than Patty Loveless. I did the correct intro and the show was superb. Randy and Patty performed for nearly two hours with many standing ovations. The next morning Randy asked if he could go walking before he ate breakfast. I met him by his room, and we walked the Galena dike around the base for about forty minutes. He was amazed at how cold it was that morning, -35°F. Later during breakfast, I told him and Patty what a great show they had done and thanked them on behalf of the military personnel for entertaining the troops around the world.

November Intercept Summary
    November 20: 1 Tu-95 Bear H

Chapter Five

GETTING BACK TO "NORMAL"
December 1988

Final Check at Eielson: Immediately following Randy's departure on the first of December I hopped on another C-12 flight for a final inspection of the barracks and alert facilities that our airmen had used at Eielson since April. I was meet at the airplane by the 343rd Fighter Wing commander, and we inspected every single space our squadron used during the deployment. What a jackass! He tried to blame our squadron for just about every little item he found during the walk-through inspection, from obvious wear and tear items to areas of the building that had not been painted in years, to a very messy open area by the barracks. Using my very best ability to keep my mouth shut, I said OK to all his comments with no intention of fixing them. Finally, I could take his BS no more and when he asked if there was any feedback from the wing's support for our deployed forces, I blew up and told him I really did not appreciate the drug sweep with the drug dogs and a no-notice barracks check on Thanksgiving day the week prior. My airmen were not able to enjoy their Thanksgiving meal as they were jacked up by the Eielson Security Police Force. No other unit on the base was subjected to that invasive inspection on Thanksgiving. I told him our airmen were more than happy to leave Eielson. He said the drug check during the Thanksgiving meal was just a random check, but I called BS on that one!

I had one final task to do while I was at Eielson. My dear wife had been air evacuated from Elmendorf to McChord Air Force Base in Washington State to have some hearing issues checked out. I had stopped at Base Operations and found that the C-141 that was carrying my wife was doing a refueling stop at Eielson before continuing on to Elmendorf and had just landed and was parked on the south end of the ramp ready to leave Eielson. Since I was scheduled to return back to Galena on the C-12, I asked the C-12 pilots if they could hold the flight for thirty minutes as I tried to meet the C-141. They agreed but they only had thirty minutes to delay or they would run into a crew duty-day limitation. Since I had no vehicle to use, I flagged down the base BITS (local mail distribution) delivery van and convinced the young airman driver to take me out to the C-141. He was very reluctant but finally agreed and drove me on the flight line right up to the waiting C-141. One of the flight engineers was preparing the aircraft, and I asked him if he would let me on the plane for just a minute or two to give a message to one of the passengers.

He agreed, and I quickly boarded the airplane, found my wife, and gave her a big hug and kiss. She was so surprised to see me, but we only had a minute to say our hellos and get an update on her hearing issues. I knew I was pushing the time limit on getting back to the C-12. The BITS driver got me back to the airplane just in time to avoid a crew rest violation. We returned to Galen and landed at about 1745 hours in the dark and a cool -45°F with snow. It was a long but productive day, and it marked the last time I made it to Eielson Air Force Base during my one-year tour at Galena.

Emergency Air Evac. At about 2000 hours on a Saturday evening while I was doing my normal walking rounds of the base, I was contacted on my portable radio that there were two very injured local residents who had been involved in a snow machine accident and needed immediate medical attention. The locals had brought the two injured people to the commercial air terminal with the hope of getting them evacuated to Fairbanks for treatment. I contacted our on duty medical technician and told him I would pick him up at the medical clinic and to bring a trauma bag. When we got to the terminal, we found two very drunk men, one with very severe injuries. They were both very combative and the most severely injured man had a compound fracture of his right femur. It was such a bad break that the bone was sticking right through his ski pants and he was bleeding profusely. My med tech immediately started work to stop the bleeding. He had to continually fight the injured person to make any progress with the injury. I saw that we would need to air evac the injured person to the closest hospital, which was in Fairbanks. I coordinated with the commercial air terminal employees, and they found a pilot and aircraft that would fly the injured person to Fairbanks. Because of the combative nature of the injured patient, I also contacted my security police section and asked them to send two SPs to the air terminal. As we continued the medical treatment, I could see that the pilot would not be able to fly the injured person to Fairbanks without medical assistance. My med tech volunteered to escort the patient to the hospital. Because of the still very combative nature of the injured person I felt that he could jeopardize the safety of the flight and my medic. I asked one of our SPs to also accompany the medic and the injured person to Fairbanks. I gave the SP specific instructions to do whatever it took, including shooting the patient if necessary, to restrain the injured person if there was any safety of flight issue. Fortunately, right after getting airborne the patient finally passed out, and the flight to Fairbanks was uneventful. Our two brave airmen returned later that evening. This event was certainly nothing I had been taught in the Base Commander's Course.

Dining In Event. One of the most legendary events in the Air Force is celebrating a unit's mission by an event called a dining in. Our entire squadron was involved in the planning, preparation, and scheduling of the dining in that occurred on December 10. Formal military dinners are a tradition in all branches of the United States armed services. In the Air Force and Navy, it is the dining in; in the Army, the regimental dinner or dining in; in the Marine Corps and Coast Guard, mess night. The dining in represents the most formal aspect of Air Force social life, and is only attended by members of a wing, unit, or other organization. The "Combat Dining In" is less formal due to the dress requirements and informal atmosphere; however, the basic rules and format of the dining in apply. Dining in ceremonies should be conducted in a tasteful, dignified manner.

Many of our customs, traditions, and procedures in the military services are traced back to the earliest warriors. The dining in is one such military tradition that has its roots in the shadows of antiquity. The pre-Christian Roman Legions probably began the dining in tradition. Roman military commanders frequently held great banquets to honor individuals and military units. These gatherings were victory celebrations where past feats were remembered, and the booty of recent conquests were paraded. The second century Viking war lords stylized the format of the victory feast. The leader took his place at the head of the table, with all others to his right and left in descending order of rank. The dining in custom was transplanted to ancient England by Roman and Viking warriors. King Arthur's Knights of the Round Table practiced a form of dining in during the sixth century.

Many early American customs and traditions were British in origin and the military was no exception. British Army and Navy units that deployed to the wilderness of America brought with them the social customs and traditions of their service to include the formal military dinner referred to as guest night. This pleasant custom provided an opportunity for officers and enlisted members to gather for an evening of good food, drinking, and fellowship. In the pioneer era of military aviation, the late General H. H. "Hap" Arnold is reported to have held famous parties called wing-dings at March Field, California in 1933, inaugurating the first of these occasions. The long association of United States Army Air Force officers with the British during World War II surely stimulated increased American interest in the dining in custom. After the war, the Air Force dining in events steadily declined in frequency until the late 1950s. The decline may have been caused by postwar demobilization, the transition of the Army Air Force to the United States Air Force, the occupation and reconstruction of Germany and Japan, and the Korean War.

Then, beginning in 1958, there was a conscious effort to rejuvenate the USAF dining in tradition. The dining in is an occasion for officers and enlisted members to meet socially at a formal military function. It enhances the esprit de corps, lightens the load of demanding day-to-day work, gives the commander an opportunity to meet socially with subordinates, and enables military members of all ranks to create bonds of friendship and better working relations through an atmosphere of fellowship. The dining in also provides a means of saying farewell to departing members and welcoming newly arrived members, as well as a forum to recognize individual and unit achievements.

Since I was the ranking officer at Galena, I was the designated president of the mess. We had the 21st Tactical Fighter Wing commander as the guest speaker this night and ended up with approximately 210 squadron members attending the dining in. We even had the two C-12 pilots that had flown the wing commander to Galena attend the dining in as a "broken" aircraft prevented them from returning to Elmendorf. We were also able to get a mandatory scramble order (MSO) status for the two alert pilots so they could join us at the big event. We found the lowest ranking member of our squadron, a one-stripe airmen from the services department, to act as the vice president of the mess. The vice president of the mess has specific duties to carry out, and our young airman did a superb job of "controlling" the mess.

One unique feature of any dining in is the use of the grog bowl. The grog bowl is an "accessory" tradition to a dining in, although not required. The purpose of the grog bowl is to "punish" any violators of the dining in protocols: speaking out of turn, using the wrong fork for the meal, not paying attention to the speaker, or responding with the wrong scripted words for a toast. Some of the specific "Rules of The Mess" from the Informal Guide to Dining-In's and Dining Out's are listed below:

The Rules of the Mess

1. Thou shalt arrive within 10 minutes of the appointed hour.
2. Thou shalt make every effort to meet all guests.
3. Thou shalt move to the mess when thee hears the chimes and remain standing until seated by the President.
4. Thou shalt not bring cocktails or lighted smoking material into the mess.
5. Thou shalt smoke only when the smoking lamp is lit.
6. Thou shalt not leave the mess whilst convened. Military protocol overrides all calls of nature.
7. Thou shalt participate in all toasts unless thyself or thy group is honored with a toast.
8. Thou shalt ensure that thy glass is always charged when toasting.
9. Thou shalt keep toasts and comments within the limits of good taste and mutual respect.
10. Degrading or insulting remarks will be frowned upon by the membership. However, good natured needling is encouraged.
11. Thou shalt not murder the Queen's English.
12. Thou shalt not open the hangar doors (talk about work).
13. Thou shalt always use the proper toasting procedures.
14. Thou shalt fall into disrepute with thy peers if the pleats of the cummerbund are not properly faced.
15. Thou shalt also be painfully regarded if the clip-on bow tie rides at an obvious list.
16. Thou shalt consume thy meal in a manner becoming gentle persons.

17. Thou shalt not laugh at ridiculously funny comments unless the President first shows approval by laughing.
18. Thou shalt express thy approval by tapping thy spoon on the table. Clapping of thy hands will not be tolerated.
19. Thou shalt not question the decision of the President.
20. When the mess adjourns, thou shalt rise and wait for the President and head table guests to leave.
21. Thou shalt enjoy thyself to the fullest.

If a grog bowl is not utilized, dining in celebrants can consider some other means of punishment for infractions. We planned ahead and at Galena had two grog bowls, one with an alcoholic beverage and a second with non-alcohol beverage. Use of the grog bowl is not to encourage alcohol consumption or public intoxication. The grog is sometimes contained in a humorous vessel, such as a clean toilet bowl, and consists of various alcoholic beverages mixed together. Our dining in team tried to make our two grog bowls as disgusting as possible, containing items such as meatballs, Tootsie Rolls, and very yellow lemonade. At Air Force dining in events, violators of the Rules of the Mess are obliged to publicly drink from a grog bowl in front of the attendees. Any member of the mess can call out violations warranting a trip to the grog bowl at any time. Members attending the dining-in bring infractions to the attention of the president by addressing the mess and raising a point of order. If the validity of the charge is questioned, members vote by tapping their spoons on the table. When the president sentences a violator to the grog bowl, the person proceeds to the bowl promptly, remembering to march and perform all proper facing (turns of 45 degrees to 180 degrees) movements. The bowl is usually located on or near the vice president's table. Upon arriving at the grog bowl, the violator does the following according the Rules of the Mess:

- An about face (180-degree turn) and salute to the President.
- Another about face and toast to the Mess: "To the Mess".
- Fills a cup with grog from the bowl.
- Drinks the cup of grog completely then inverts the cup on top of the head to ensure the cup is empty.
- Does an about face, replaces the cup, about faces again, salutes the President and returns to their seat.

During our dining in the airmen had really studied the Rule of the Mess, and I was worried that nobody would make a trip to one of our very nasty grog bowls. I then initiated a couple of violations, that the president of the mess has authority over, by sending two of my captains and a two of our master sergeants to the grog bowl. That really opened the flood gates and many of our airmen and officers made trips to the dreaded grog bowl. The trip to the grog bowl is meant to provide a relaxing humorous social encounter and build team identity for the unit's mission. The entire evening was a huge success and the festivities really brought our squadron together following the long alert deployment to Eielson Air Force Base.

Intercepting the Bad Guys. Again, our primary mission at Galena was to intercept and divert any Soviet aircraft incursions into U.S. air space. Following a couple of failed practice alert scrambles, the squadron finally got all the bugs worked out of our newly renovated combat alert cell, and we were ready to carry out our mission in the middle of December. Since I had flown many intercept missions in the F-106 a few years earlier, I was completely tuned in with the alert process and was elated that we were finally ready to fulfill our mission. That alert scrabble happened on December 15, 1988. Early that morning the CAC was given a Warning Order (an intelligence tip that an activity is excepted to happen) from the intelligence folks at the Alaskan NORAD Region Operations Center to expect two Soviet Tu-95 H Bear Bombers to attempt to enter the AIDZ later in the day. The normal procedures for these intercepts was to have the Airborne Warning and Control (AWACs) aircraft in the sky to capture precise tacking information on the Soviet bombers. In addition, we usually had a KC-135 tanker aircraft airborne for aerial refueling of the F-15s.

The F-15C models that we had on alert were state-of-the-art fighters. They were a rather large aircraft, with a forty-three-foot wingspan and an overall length of nearly sixty-four feet. The aircraft was piloted by a single officer and had the capability of reaching a speed of Mach 2.5 (two and a half times the speed of sound) with a max operating ceiling of 65,000 feet. It could carry a wide variety of weapons including the AIM-7 radar missile, the AIM-9 infrared missile, and the new long-range AIM-120 AMRAAM missile. The AMRAAM is a very high-tech missile that gives our pilots the beyond visual range (BVR) air-to-air missile capability. It also carried an internal 20 mm M-61AI Vulcan 6-barrel rotary cannon that held 940 rounds. Our pilots also carried a handheld 35 mm camera in the cockpit to record critical information on each intercepted Soviet aircraft. The F-15 pilots changed out every seven days, usually on a Thursday, and every two weeks they would fly to Galena on a C-12 aircraft.

The following week the pilots would change out by bringing two different alert aircraft to Galena and change out with not only the pilots on alert but also change out the alert aircraft on the same day.

The Airborne Warning and Control (AWACS) E-3 aircraft were highly modified Boeing 707 aircraft that had a rotating radar dome attached the top of the aircraft. The Alaska AWACs aircraft were part of the 962nd Airborne Air Control Squadron based at Elmendorf. The E-3 had a flight crew of four and a mission crew of thirteen to nineteen radar operators. The AWACs represented a major increase in tracking capabilities as it used a pulse Doppler radar which allowed it to track targets normally lost in ground clutter. The radar range on the AWACs aircraft was approximately 220 nautical miles. The aircraft was 136 feet long, with a wingspan of 131 feet. It could carry a fuel load of 200,000 pounds and could stay airborne for extended periods of time.

The KC-135 tankers were also critical to our intercept mission. The KC-135 was an aerial refueling aircraft that provided extended on station time for our F-15s. The Alaska KC-135s were part of the Alaska Air National Guard's 168th Air Refueling Squadron based at Eielson Air Force Banks in Fairbanks. The KC-135, a derivative of the Boeing 707, was the mainstay of the Air Force refueling mission. I had many refueling missions with the KC-135 when I flew the F-106 out of McChord Air Force Base. The KC-135 was 136 feet long with a wingspan of 131 feet and could carry nearly 200,000 pounds of fuel. The tankers that supported our intercept missions were on twenty-four-hour alert status.

The Tupolev Tu-95 G and H model Bear bombers were the aircraft that Soviets used to test the U.S. resolve and our ability to intercept their aircraft flying toward and near U.S. and Canadian airspace. The Bear was a very large, propeller-driven aircraft that was a quite capable platform to launch cruise missiles. The aircraft was 151 feet long, with a wingspan of 164 feet. It had a loaded weight of 376,00 pounds, could cruise at 340 miles per hour, and had an unrefueled range of 9,400 miles. The crew consistent of six to seven personnel, including a tail gunner; the tail gunner had two 23 mm radar-controlled guns at his station.

The Soviets used these bomber flights near Alaska to train their aircrews to fly over the arctic region and launch cruise missiles into Canada and the lower forty-eight states. In these training operations the Soviets would launch Bear bombers about every fourteen days on nuclear strike training missions which approached as close as fifty miles of the Alaska coast without violating U.S. airspace. The Soviets also launched a number of electronic intelligence missions and polar ice survey missions which our U.S. forces also intercepted.

So, this was the lineup for the intercept mission on December 15. The alert set up at Galena was very similar to most fighter alert facilities supporting the NORAD mission. Depending on the intelligence data and long-range radar data, the F-15 crews had different levels of alert status including the following levels:

- **Suit up:** The F-15 pilots were notified that they should be in their flying gear and expect a higher alert status momentarily.

- **Battle Stations:** This status required the pilots to be suited up and strapped in the cockpit before starting engines.

- **Runway Alert:** The pilots would start their engines and taxi to a takeoff position near the runway waiting for a takeoff order from the ROCC.

- **Airborne Orders:** This status provided several minutes of advance notice to the pilots and designated an exact takeoff time to be airborne.

- **Scramble:** Actual no notice takeoff to intercept a Soviet aircraft before it reached the ADIZ.

During my intercept flying days in the lower forty-eight states out of McChord Air Force Base, Washington and Kingsley Field, Oregon, we had a fireman's pole to quickly slide down from our sleeping quarters right into the hangar next to our aircraft. Galena also had the fireman's pole, but it was never used during my year at Galena as the alert pilots always had prior warning from NORAD of any potential intercept activity.

Takeoff weights for the F-15Cs flying out of Galena with two wing tanks, two conformal fuel tanks and eight missiles was around 67,000 pounds. With a short, 7,250-foot runway, (the shortest operational fighter runway in the Air Force) afterburner takeoffs were mandatory. Once airborne, the F-15 pilots contacted the ROCC and were then vectored to the tanker before the intercept or directly to the Soviet aircraft. The AWACS directed the F-15 to about twenty-five miles from the bomber, at which time the F-15 completed the intercept using its on board radar. The approach was usually from above or below the bomber to negate any contrails the F-15 may be forming. If the fighters were intercepting two bombers in formation, they would almost always intercept the trailing bomber. They normally moved no closer than 500 feet from the bomber. The first fighter closed to about 500 feet while the second fighter remained back in a supporting defensive position in case the bomber tried any defensive moves.

Chapter Five

Once in position the lead fighter took photos of the bomber with particular attention to the bomber's tail number. For many years we knew that the Soviets would change the tail numbers of each bomber with the hope of confusing the U.S. intelligence about the number of bombers in their inventory. However, through some ingenious intelligence work we realized that even though the Soviets changed the tail numbers they failed to change the numbers on the bomb bay doors of the bombers. So, our fighters routinely dropped below the bomber and photographed the bomb bay door numbers giving us an actual count on the different bombers.

The intercept on December 15 involved two Tu-95 G bombers that were intercepted about one hundred miles north of Point Barrow, Alaska and only 600 miles from the North Pole. The two bombers came within eighty miles of the Alaska coastline. During this entire year, the Soviet bombers did not – or could not – track our aircraft with their radar-controlled tail guns. Many times, the bomber crews waved to our pilots. However, our pilots were directed to never casually engage the crews in the bombers. The pilots did observe some funny things at times from the bomber tail gunner: the Soviet tail gunners took great pride in either holding up a Coca-Cola can or spreading out the centerfold of a Playboy magazine to show that they did have some of the same western "culture" as the Americans.

Upon return to Galena our F-15 fighters had another challenge to face landing on our short runway. If the runway condition reading (RCR) was anything lower than a nine (wet) reading they would engage the approach-end runway barrier (cable) on runway 07/25 to ensure a safe landing and roll out. The barrier arresting system at Galena consisted of large, one-inch steel cables stretched across each of the approach ends of the main runway. The cables in turn were connected to large, in ground, B-52 aircraft brake components to gradually play out the cable once the tail hook of the fighter hooked the cable. Our Galena barrier maintenance crews were some of the very best in the Air Force as they had to immediately get the first aircraft out of the arresting cable and reset the cable for the second aircraft to also engage the cable. Winter flying in and out of Galena was a real challenge. Whenever the runway at Galena become too slick for normal operations, we reverted to a mandatory scramble order (MSO) status where only real-world intercept missions would be allowed to use the slick runway. If the aircraft were airborne when runway conditions deteriorated to an unsafe RCR reading, the aircraft, with tanker support, would recover at either Eielson's 14,500-foot runway or Elmendorf's 10,000-foot runway.

Alcohol, Alcohol, Alcohol. I had been warned by the 21st Tactical Fighter Wing social actions office prior to my arrival at Galena that all of the Alaska remote sites, especially Galena and King Salmon, had an intense alcohol abuse problem with the military and civilian staff. I inherited seven very serious alcohol abuse cases involving members of my squadron. Three of the abusers were already on temporary duty (TDY) to Elmendorf undergoing the intense four-week Substance Abuse Education and Recovery (SABER) Program. In 1988 it was Air Force policy that if a person self-identified as having a substance abuse problem they could voluntarily enter the SABER program without any military judicial consequences. So here I had three valuable team members away for four weeks going through an intense alcohol rehab program. It did not matter that the remaining team members had to pick up the slack of those members that were TDY to Elmendorf. Nevertheless, this was the policy and each of the three members were making satisfactory progress with their rehab. I even stopped in to visit the three when I was back at Elmendorf on temporary duty doing some 21st TFW wing business.

The second two individuals I was dealing with were altogether different situations. We had a very talented technical sergeant in the civil engineer branch who was a recent graduate of the SABER program. He obviously did not do too well in the program and was back at Galena still drinking and causing problems in our squadron. He had punched a subordinate in the face over a very minor incident and was once again found drunk at his duty station one Monday morning. I called him in to my office to counsel him on his behavior and told him that he would need to retake the SABER program if he intended to stay in the Air Force. He was very respectful to me and said he would do better in the future. We then started coordinating to get him back to Elmendorf for follow-on SABER treatment. I also restricted him from driving any vehicle during his remaining six months at Galena. The following Saturday evening at about 2230 hours I got a call from the front gate shack into Galena that the sergeant and a civilian Air Force employee had returned to base in a very inebriated state. The civilian, driving an Air Force pickup, just blasted through the front gate ignoring the orders from the gate guard to stop. The civilian was bringing the sergeant back on base and did not want us to know that the sergeant was drunk on his ass. The civilian was also very drunk, but we were not able to stop him to get a blood alcohol test (BAT) on him because he once again raced through the gate on his way out of the base. The gate guard certainly had a good facial recognition on him as he left the base. I guess he figured if we couldn't do a BAT on him, he would be home free.

We knew that he would return to the base on Monday morning for his daily shift, and I had the security police waiting for him. He was apprehended at the gate and was brought to my office for counseling. He was very surly and really didn't think he did anything wrong. Since I needed to consult with the civilian personnel office at Elmendorf, I had limited options for him. I did restrict him from driving either a military or civilian vehicle on base for six months. For the sergeant, I counseled him again and advised him that I was punishing him with an Article 15, busting him down one rank to staff sergeant and restricting him from driving on Galena until the time he was transferred to his new base. Unlike a formal court-martial, an Article 15 is a form of non-judicial discipline carried out by commanders. An Article 15 does not result in a criminal record and may not affect a service member's record in the military; in other words, an article 15 is administrative punishment, usually for minor misconduct. There are some service branch-specific variations, but if misconduct is established, possible administrative punishments include loss of rank, loss of pay, restrictions, and extra punishment duties.

The next two drinking incidents involved individuals who were supporting our deployed alert mission at Eielson. One of the individuals was drunk and drove one of the rental cars the squadron was using during the deployment into one of the lakes on Eielson. He really did not have any excuses as his BAT was .10 above the .08 DUI level for Alaska. I also busted him from staff sergeant to sergeant and restricted his driving on base for the remainder of his tour at Galena.

The second, but certainly not the last individual, I had to deal with during the year at Galena involved an individual in the fire department. It seemed that he also was a recent graduate of the SABER program and his supervisor had given him a four-day pass to Elmendorf for a reason that escapes me. I asked his supervisor if he would be OK going to Elmendorf on a four-day pass and he said, "No, he plans on drinking and partying the whole time he is there."

I immediately canceled the four-day pass and had the sergeant in my office the next morning laying out the criteria for him to remain a sergeant and not get busted for his drinking problem. This was not the last time I had an issue with the fire department supervisor who issued the four-day pass.

December Intercept Summary
    December 15: 2 Tu-95 Bear Gs
    December 28: 1 Bear Tu-95 Bear G

Chapter Six
ARCTIC FREEZE
January 1989

The very best way to describe January 1989 at Galena was COLD. Most of the stories related to this month deal with the record-breaking cold spell that hit our arctic base and nearly brought us to our knees. However, the cold was not the only issue I dealt with. Once again, the 10-percenters took up a great deal of my time as I also dealt with those issues and the impact of the record-breaking cold temperatures.

Dormitory Damage. As part of my normal "leadership by wandering around" routine I tried to visit all the departments that were working on New Year's Day 1989. The morale was pretty high at just about every stop I made. The senior enlisted dorms were relatively quiet; however, the junior enlisted dorm, building 1872, was very loud and boisterous by the time I arrived at around 2200 hours. As soon as I walked in the building, I noticed a great deal of damage had been done to the ceiling tiles and lights in the main hallway of the first floor. I was none too pleased at what I observed and asked everyone that I saw in the building if they knew anything about the damage and who may have done the damage. Not a word was shared about the damage, but I was bound and determined to get to the bottom on the nearly $5,000 dollars in damage to the dorm.

Early the next morning I held a special staff meeting with all my senior enlisted personnel and all my officers to try to get to the bottom of the damage to the dorm. I made a command decision and advised all my senior leaders to tell the squadron that there would be no Date Estimated Return from Overseas (DEROS) flights until the culprits were found and held accountable for the damage. A DEROS flight is the first of many flights an individual took while leaving Galena and travelling to their next assignment. The damage happened on January 1, my decision to hold all DEROS flights was on January 2, and the next scheduled DEROS flight was not until January 10. I felt confident that we would have an answer on who did the damage before the 10th. In spite of my "pressure" on the squadron no one came forward claiming responsibility for the damaged dorm.

On January 10 my operations officer called me and told me he had three folks at base operations that were expecting to leave Galena on their contract DEROS flight in about one hour. He wanted to confirm that my edict on no departures was still in place. I drove to base operations and told the three individuals that they would not be leaving Galena that day and would not be leaving Galena period, until the damage at the dorm was cleared up. I got the standard push back from them that they would call the inspector general, their congressman, and anyone else they could think of. Fortunately, I had planned ahead and knew the exact dates the three individuals had arrived at Galena the year before. None of them arrived at Galena before January 25, 1988. Letting someone leave a remote assignment before the actual one-year anniversary was a privilege the commander could grant, if requested. I also had the name and the phone number of the inspector general at Elmendorf and told them to feel free to call and complain. They were none too happy but did return to the visiting enlisted dorm and were ready to get another date for their DEROS flight. None of them contacted the IG or their senator.

I guess I may be a little warped but thought I would have some fun with the dorm damage investigation. I was notified on January 3 that the list for the newly promoted staff sergeants was released. I had my first sergeant call all seven squadron members that were selected for promotion to come to my office at 0730 hours the next morning. If any questions arose about the meeting in the commander's office, I wanted everyone to believe it had to do with the damaged dorm. As I came to the office, I saw seven very long and concerned faces outside my office. I made a point of having our head of security police outside the office as well, and I asked the seven to come into the office. "I guess you know why you are all here this morning," I started. You could feel the tension in the room explode when I told them, "You are here today…pause…because each of you have been selected for promotion to staff sergeant, congratulations."

The word got out rather quickly that I was still investigating the dorm damage and all DEROS flights were on hold this month until further notice. Between January 4 and January 16, I had nine end-of-tour (EOT) out-briefs. Really the only question from all nine was "Will I be able to DEROS on time or are you going to hold everyone indefinitely until the culprit that caused the damage came forward?" My answer was very vague on purpose. I told them I would, one by one, eliminate those that did not cause the damage and consider approving individual DEROS flights. I did remind them that leaving before the end of their DEROS month was my decision alone.

The next step in the process of finding the culprit who damaged the dorm was to put the day room and pool room off limits. A flurry of complaints about that decision piled up with my first sergeant, who was superb during this entire process. He kept the pressure on the airmen from the dorm and relayed all of my decisions directly to the airmen so there would be no confusion as to my plans to solve this mystery. I was really starting to think we would never solve this mystery as the folks from the dorm seemed to have agreed to a code of silence. But the threat of no DEROS flights until someone confessed finally paid off.

Early on the morning of January 14, I was met at my office at 0700 hours by two very young airman who wanted to talk to me about the damage in the dorm. Neither one of them had a DEROS flight scheduled in January but did not want to hold up any of their friends' DEROS flights. So, they both confessed that in a drunken stupor they had damaged the ceiling and ceiling lights in the dorm. I thanked them for their honesty but told them that each would have a Letter of Reprimand put into their records and that neither of them could leave Galena until the damage was repaired. Luckily for them they were both assigned to the civil engineering unit and they had access to materials and supplies to repair the damage. The damaged ceiling and lights were repaired by the end of January, and I never had another problem with those two airmen. Word got out that the commander would follow through on any disciplinary actions he thought were appropriate. I usually managed to go to base operations and say farewell to all the airmen that were completing their remote tours during any given month. The only problem for those leaving in January was that we had record-breaking cold temperatures which closed the Galena runway to all flight operations, both civilian and military, for several days. The runway and regular flight operations did not start again until February 1.

Dealing with the COLD. As I said in the introduction to this chapter, we experienced the coldest weather ever recorded at Galena during the month of January 1989. These low temperatures broke the all-time coldest day record that was set in 1961. To give you an idea of what we dealt with, the temperature range between January 15 to January 27 was a high of -36°F to a low of -70°F. From January 20 through January 27 the temperature range was from -55°F degrees to -70°F degrees. We certainly had many issues with those temperatures and, there were many actions we had to take to "save" Galena from the deep freeze.

Day-to-Day Operations. The first indication of impending disaster came on January 12 when I got a call during lunch that the water pipes in the Galena Consolidated Club had burst and there was a great deal of water flowing out the front door and immediately freezing into a pool the size of a skating rink. The civil engineer staff were on the problem in less than thirty minutes and turned off the water to the building. The cleanup began in earnest, but it was still a huge mess. The weather that morning was -43°F, but not an unusual temperature for Galena in January. I consulted with my Air Force meteorologists and they said that all of Alaska was sitting under a huge high-pressure area and the temperatures were forecast to be well below freezing for the next two to three weeks.

The temperatures continued to drop, and on January 16 we hit -64°F and many problems followed. Pipes in our communications building, the combat alert cell, the junior enlisted dorm, base operations, and the air traffic control tower all had frozen and many had broken. On top of that we were getting calls that almost all of the vehicles that were kept outside would not start or the wheel bearings were frozen solid. We had a problem on our hands, and I immediately called an emergency staff meeting with all my officers and senior enlisted leaders to come up with a plan to combat these extremely low temperatures. We had no preexisting checklists to deal with this freezing weather, so we were kind of flying by the seat of our pants trying to come up with a plan.

While we were convened in the staff meeting, I was notified that the dining hall, the transportation office, and the power plant office all had frozen pipes. Additionally, we were notified by the civilian air terminal that they also had frozen pipes and the Mark Air refueling station was frozen solid; there would be no more commercial air traffic into or out of Galena until further notice. I coordinated with the 21st Wing and the Alaska NORAD Region to recommend that the F-15s be placed on mandatory scramble order (MSO) status because living quarters at the CAC also had frozen pipes. That turned out to be a very wise decision because I got a call just after the staff meeting to meet my operations officer on the runway for a problem he had detected.

My staff vehicle, a 1984 Chevy Suburban, barely started but I was able to meet the ops officer on the runway and what I found was very scary. He told me to turn off the vehicle and just listen. We heard these fairly loud popping sounds (like popcorn popping) coming from the runway and saw stones the size of quarters come flying out of the newly paved runway. The moisture in the pavement was freezing, and the runway was literally coming apart. Large horizontal cracks appeared that were six to eight feet long, two to three inches wide, and five to eight inches deep. We were in "deep shit" at Galena. The temperature was -63°F. And that was just the start of an incredible sixteen days of trying to save Galena. As soon as I returned to the office, I notified the wing commander at Elmendorf that we had a major problem with the runway, that most of our buildings were freezing up, and we could soon be non-operational. The commander told me to get him a list of the things we needed to keep Galena open. I reconvened my staff and went around the room asking each functional leader what they needed to save Galena. Before leaving my command vehicle outside again I asked my first sergeant to find me a young enlisted airman that would keep my vehicle running by driving around base and be ready whenever I needed the vehicle. It turned out to be another wise decision as most vehicles left outside, even with the engines running, were experiencing frozen wheel bearings.

The vehicle issue became critical as we needed to decide which vehicles were absolutely mandatory for the current cold weather emergency and what vehicles could be sacrificed. The list was very inclusive: all of the fire department vehicles, four security police vehicles, numerous civil engineering vehicles, some aerospace ground equipment, the base water truck (which turned out to be a prophetic decision), and, of course, the commander's vehicle. Where we could, we assigned drivers to keep some of these critical vehicles moving, and the rest were stored in the very large birchwood hangar with a few portable heaters running nonstop. That seemed like a good plan until we got word that our vehicle refueling building and main refueling pumps on base were now frozen solid. We now had no way to refuel our running vehicles.

Friday, January 20 was another very cold -60°F day. We started off again with another cold-related crisis. Our missile storage build was now frozen, the headquarters building was frozen, and the metal handles on the headquarters front doors were frozen – they just broke off as we tried to open the building. We also learned a valuable lesson that day about very cold temperatures and the new technology in the newer vehicles; if a post-1985 vehicle stalled and the maintenance experts closed the hood, all the plastic components near the engine or the various drive belts simply shattered when the hood was dropped down. I gave new directions to the transportation folks to just leave the dead vehicles where they are unless they were causing a traffic issue.

Just when we didn't think anything could get any worse, I was informed that the diesel fuel that powered our generators for the base was starting to gel, at -54°F, and couldn't flow into the generators. This gelling process had caused a fuel spill at the power plant as the fuel was backing up in the transfer lines from the external fuel tanks to the internal storage tanks of the power plant. We were now down to just two of the four generators that were providing power to the base. I called an emergency meeting with the power plant operators and tried to get a solution to this very critical problem. Fortunately, one of the very senior staff at the power plant said he thought if we heated the diesel fuel to above -54°F as it was pumped from the external storage tanks to the generators we may be able to keep the fuel flowing. I called back to Elmendorf and requested ten Herman Nelson 250,000 BTU/hour gas-powered portable heaters be sent immediately to Galena or we may have to abandon the base. The wing got right on the problem and dispatched a C-130 with the heaters. I had to temporarily open the runway for the C-130 to land, but we had our heaters in less than six hours and we solved the freezing fuel problem. The civil engineer experts placed heaters near the output fuel line from the large external fuel tanks and near the intake fuel line to the generators and the heated diesel fuel began to flow into the generators, saving the day and the base.

Just as we were dealing with the power plant issue, we got a notification from the alert hangar that all of the fire alarms activated at the same time. Fortunately, the fire suppression system was frozen so no water was sprayed on the two alert F-15s. I immediately ordered the F-15s to evacuate the alert hangar and taxi away from the building until we could determine the cause of the activation of the fire alarms. My civil engineer was convinced that the fire alarm activation was related to the cold weather and that they would have it fixed in short order. We ordered the crews to keep engines running on the two F-15s until we could solve the problem. It was such an amazing site as the two aircraft started their engines and taxied out, they were both sending huge contrails, just like you see out of the back of aircraft in the sky. The temperature during this crisis was -60°F.

After this fire drill was under control, I picked up my civil engineer officer to do a tour of the base and get an idea of the amount of damage we were sustaining from this cold weather. We stopped in front of an old barracks building that had been closed for the past year awaiting demolition. The CE officer said, "You need to hear and see this incredible sight." As we entered the building, I could hear a very loud cracking sound followed a few minutes later by another loud cracking sound. It seemed that this building was freezing as we walked through it and the noises we were hearing were the abandoned toilets in each room exploding as the frozen water expanded and cracked the toilet water tanks.

More freezing issues occurred as one of the two main boilers that provided heat to the base was down. We were without full capacity heat for nearly forty minutes and that really exacerbated the freezing of additional buildings on Galena. I called for another emergency staff meeting and for the first time addressed the possibility that we would have to abandon Galena if we didn't get any relief from the cold. I asked each functional area leader to designate the emergency personnel that we would leave behind if we needed to abandon the base. We came up with fifty positions. The personnel folks made travel orders for the remaining 270 personnel that we would evacuate to Elmendorf if necessary. The temperature was now down to -70°F, a new low record for Galena. With a five-mile-per-hour northwest wind, the wind chill temperature equated to -88°F. More damage to our birchwood hangar occurred when the main civil engineering supply room was flooded from a broken pipe. More vehicles died in the cold, transmissions froze, and frozen wheel bearings kept the majority of the motor pool fleet grounded. Our one and only road grader was basically destroyed when the roads and ground crews were trying to level one of the main roads by the power plant. The cold weather caused the nearly twelve-inch-wide and two-inch-thick steel frame on the grader's blade to crack completely in half. It would be some time before we could get a replacement. Still, I held off directing an evacuation of Galena.

Brim Frost 1989. In the midst of the coldest winter on record in Alaska, the Alaskan Air Command was hosting the every-other-year military exercise, Brim Frost 89. The exercise that year included 26,000 military personnel, 120 aircraft, and 1,000 vehicles from both the U.S. and Canada. The purpose of the exercise was to test personnel and equipment in freezing temperatures. That certainly was a valid test in 1989. The exercise included the U.S. Army, Air Force, Marines, Coast Guard, National Guard, and Reserve units as well as Canadian ground and air forces. Galena's role in the exercise was to provide support to mostly flying operations but also support to a contingent of Eskimo Scouts. We were expected to support an additional two alert F-15 aircraft and six adversary aircraft for the mock air battles.

The Eskimo Scouts were a very unique and legendary organization. A highly specialized unit of the Alaska National Guard, the scouts used centuries-old Arctic survival skills to conduct surveillance and reconnaissance missions on U.S. terrain that lies at the frigid back door of the Soviet Union. Their place in the U.S. military was so unique that they sometimes trained the Army's elite special operations forces sent up from the lower forty-eight states. The 1,500 scouts were drawn from ninety-one isolated native villages along the west coast and interior of Alaska. Most scouts were subsistence hunters and fishermen who spent the spring and summer pursuing the walrus, seals, whales, caribou, moose, and fish that provided their families with food, clothing, and income. Using the same skills for surviving in one of the world's least hospitable places, they led snowshoe marches across miles of roadless white tundra while packing rifles and survival gear that included body-warming chunks of whale blubber, muktuk, and strips of dried fish. They navigated the darkness of the treeless Arctic desert, relying more on instinct than a compass, in a place where an unexpected blizzard could turn a military exercise into a real life-or-death survival test. Most of their military combat training focused on unconventional warfare and guerrilla tactics, and the best earned the National Guard's coveted "Arctic Warrior" badge. But combat training was secondary to the scouts' mission of watching for any air, sea – and sometimes land – operations of military activity from the Soviet mainland just thirty-seven miles west of the coast of Alaska. So we also had to be ready to support these additional forces during our crisis with the cold weather.

Time to call it quits. With all the cold weather issues we were facing, I finally called the wing commander and requested we be exempt from supporting the Brim Frost 89 exercise activities. I told him our main goal at that point was to keep Galena operating, but we would try to do whatever we could to support the exercise. Since our runaway was unusable, the flying portion of the exercise from Galena was cancelled. The latest challenge was the power plant; even with the fuel now flowing to the generators, we were having a difficult time keeping up with the heat demand from the cold weather. I made a command decision to turn off all heat to all non-essential facilities to try to save the essential buildings on base. The weather that day was -70°F again.

I got a call later that day from the Alaskan Air Command headquarters to check on the Eskimo Scout unit that was camped right outside our base. Since my vehicle was being driven twenty-four hours a day, seven days a week, it was all ready to go as I ventured off base to a point just northeast of the dike. I saw the eight tents of the Eskimo Scouts but did not see any smoke from campfires coming from the encampment. I stuck my head into one of the tents that looked like the leader's tent and found a young Eskimo sergeant shivering in the tent with no apparent heat. I asked him how his team was doing, and he told me they were freezing their asses off. It was -70°F outside and probably -40°F inside the tent. I then got on the radio and relayed back to Elmendorf what I had seen and that the Scout team was in definite trouble. I recommended that they be evacuated back to Galena for their survival. After a very short delay the U.S. Army Brim Frost 89 coordinator at Elmendorf agreed, and I relayed the info back to the Eskimo Scout sergeant. I offered to come back with vehicles to pick them up, but they refused and said they would march the four miles back to base. While driving back to the base I talked to our services officer about a place for the Scouts to stay. Since our flying activity in support of Brim Frost had been cancelled, we did have some rooms available for the twenty-five Scouts in one of our visiting officer's quarters. As soon as the scouts returned to base, I offered the visiting officer's quarters to them, but they refused and instead preferred to stay in the closed barracks that had all the broken toilets. They would sleep on the floor and make the best of the situation. I did not turn them down with their request to stay in the abandoned barracks but did direct them to the dining hall for a hot meal. They certainly enjoyed that meal. By the next morning the Alaska National Guard had sent a C-130 to Galena (I temporarily opened the runway again) to transport the Scouts back to their home base in Bethels, Alaska. What a tremendously tough and dedicated group of soldiers.

City in Distress. Because of the myriad cold weather issues we were facing, I set up two standing staff meetings for each day at 0800 hours and 1600 hours to address our problems and solutions to this arctic deep freeze. On January 27 the temperatures again dropped to -70°F, and I was contemplating abandoning Galena for the safety of our squadron personnel. There were so many issues that we barely had a handle on. During the morning staff meeting my administration clerk told me that I had a very urgent call from the mayor of Galena and that he needed to talk to me as soon as possible. I turned the staff meeting over to my vice commander and went to my office to take the call from the mayor. The mayor was very distressed and told me that the city water supply had completely frozen, the city boilers were down, and his people were running out of food as there had been no resupply flights for the past four days. He said some of the folks in town had resorted to eating some of their sled dogs as there was no food in town. He asked if our squadron could do anything to help. Wow, just another challenge on the list of things we needed to do. I returned to the staff meeting, explained the dilemma with the city, and asked my civil engineering officer and the chief of security police to accompany me to town to meet with the mayor and devise a plan to help save the city.

Our meeting with the mayor was short and to the point. The city needed water, food, and a heat source if they were to survive this arctic blast. I ensured the mayor that we would do all we could to help but that were also dealing with the cold weather crisis on base. As soon as I got back to base, I reconvened the staff and explained the dilemma the city was facing and asked for suggestions on what we could do with our limited resources and our own efforts to deal with the cold as well. The answers came immediately from the staff. The first sergeant and supply officer said they would scrounge the base for any excess wood products to use as fuel for the city, our civil engineer officer said that, despite the cold, we did have a source of running water that he could use to fill the 7,500-gallon water tank truck and deliver it to the city. Remember the wise choice we made to keep the base water truck running during the cold? And finally, our services officer said we could donate about fifteen cases of meals-ready-to-eat (MREs) to the city.

First, we had to gather the wood to take to town to use as fuel. The supply folks had a huge pile of wooden pallets that were loaded on the three trucks that we had kept running. Our services officer decided that we could donate all the ranch oak furniture that was left in the barracks that was going to be demolished, and he rounded up the fifteen cases of MREs to load on a truck for the city. Our first sergeant also contacted each work section on the base to request volunteers to help with the wood roundup, asking each section to find any usable wood that could be donated to the city. After the scrounging was complete, we were able to find nearly seven loads of wood for the city.

The base chaplain agreed to lead the team effort to deliver the food, water, and wood to the city. By the time we had everything rounded up, it was too late in the day to deliver the supplies. On January 29 it had warmed up to -48°F, and the resupply effort was on. In addition to all the food, water, and wood, our very generous squadron members rounded up about fifteen bags of extra warm clothing to donate to the city. The caravan to the city took three trips as we did not have seven pickups that were running that day. I drove into town for the first drop off, and the city officials were amazed at what we had brought to them. The water distribution process was amazing – we parked our water truck inside one of the large school storage buildings and had a few portable gas-powered heaters set up inside the storage area. The local residents came to the water truck and filled as many individual containers as they could carry to bring water back to their homes. They had a real challenge getting the water back home before it froze. We off-loaded the firewood in a central location in town and watched as the locals gathered as much as they could carry back to their homes. We gave the food to the mayor and told him to distribute it the best way he thought. We did the same with the clothes. All in all, we did a superb job, and the city was very pleased with our help.

Cold Issues Continue. On January 27 I received word that the Air Force-operated runway approach control radar on the airfield was frozen. The problems continued with the further deterioration of west end of the runway. I put a visual flight rules (VFR) restriction on military flights into Galena. I did open the runway for only emergency C-130-type aircraft to land and operate on the runway. I designated a team of operations folks to conduct foreign object damage (FOD) inspections of the runway three times per day and to pick up all of the loose stones that they found. It was a pain in the ass for them but at least we could get the C-130s into Galena with some much-needed supplies. In addition, I coordinated to get a runway inspection team from the Air Force Headquarters to come to Galena on February 2 to inspect the runway and determine our options for further operations. It did warm up to -38°F on January 31, so it seemed that the cold spell was breaking. I was getting concerned about my squadron members as they were working outside without their parkas as it "warmed" up.

January Intercept Summary
    Zero intercepts

Chapter Seven
PICKING UP THE PIECES
February 1989

"**Heat**" **Wave. February finally arrived, and the temperature** rapidly rose to a "balmy" -30°F with warmer temps forecast for the rest of the first week of the month. Now warmer temperatures certainly did not mean our arctic freeze problems were over…far from it. As temperatures began to rise, we found hundreds of additional pipes that had frozen during January, and now that they thawed a bit the water leaks were biblical. With all the plumbing issues we were facing, I requested two more highly qualified plumbers be sent from Elmendorf on a temporary basis just so we could stay ahead of the water leaks.

Hospital Buffoonery. We had a young airman plumber that was not doing very well at Galena during his two months on station (he arrived at Galena in mid-November). He was showing signs of mental illness and was always crying to his supervisor that he could not be assigned to Galena because he missed his family. He was rather useless to us as we tried to deal with all the cold weather issues that required a degree of plumbing expertise. On the very first flight out of Galena after the runway was open for commercial traffic, we sent him back to Elmendorf for a mental health evaluation. As our pace of dealing with the cold weather crisis increased, we just forgot about him being back at Elmendorf until we realized we were short a plumber. I called the hospital to get a status on the young airman, and they told me he was doing OK but he shouldn't be assigned to a remote site. I requested that the hospital document the diagnosis in an official letter so that we could requisition another plumber for Galena. The hospital refused, telling me that an official letter could hurt his career. So on the very first next commercial flight to Galena from Anchorage I met the airplane because there were several other squadron members on the flight that we wanted to put to work immediately. As the young airman plumber got off the aircraft, he walked directly to me and said in a very matter of fact way that if he spent another night at Galena, he would kill himself. Great, just what we needed. I called the first sergeant to the terminal and told him to put the plumber on the next flight back to Anchorage (one hour later), to accompany him on the flight, and to escort him directly to the hospital. That was the last time we saw him at Galena; we packed up all of his personal belongings and sent them to the Elmendorf hospital. That was also the last time I would trust the wisdom of the Elmendorf hospital. They confirmed that mistrust later in the year with another very weird mental health evaluation with an older sergeant from our squadron.

New Leadership Ideas. One of the many challenges of being on a one-year remote assignment was not being able to share key events in life with family. In close coordination with a couple of key staff members, we came up with two monthly events that would at least, to a small degree, make up for celebrating key events without family. First on the agenda was a recommendation from our food service superintendent. Now this was not just an ordinary food superintendent. This very talented master sergeant would be recognized in September as one of the top food superintendents in all of the Air Force by winning the Hennessey Award competition. The Hennessey awards recognize outstanding food service individuals throughout the world. Our own Galena food service superintendent was the winner of the best small base food superintendent award. He came up with a brilliant plan for the commander, me, to have a monthly birthday meal with everyone in the squadron who was celebrating their birthday in any given month.

The birthday meals were held in the commander's dining room adjacent to the main dining room. The food experts from the dining hall prepared superb birthday meals with whatever key ingredients they could get flown into Galena. Sometimes it was steak, sometimes shrimp, and sometimes a delicious lasagna dinner with all the trimmings. The meals were delivered to each individual by the food staff just like restaurant service and were topped off with a large sheet cake that honored all the birthday people. One of the things I did after each meal was to ask the attendees to tell us about where they grew up, a little about their family background, and what activity they would be doing for their birthday if they were home with their families. We received very good feedback from the squadron on this event.

Another monthly event I hosted was a morning breakfast, again in the commander's dining room, honoring all the newly selected airmen for promotion during any given month. We all would just grab our breakfast from the food line and convene in the commander's dining room. My first sergeant would have the new stripes for the airmen, and I would individually hand out the stripes, say a few words about each person, and recognize the hard work they were doing in our squadron. It was a very simple event, but it recognized the tremendous efforts each had done to earn their promotions.

Runway Repair Project. One of the main issues from January's incredibly cold weather was the condition of the runway and what needed to be done to repair it. I certainly did not want to send the alert fighters to another base while the runway was being repaired. We did have some civil engineering experts from Air Force Headquarters come and "look" at the runway early in February. They were all amazed at the extent of the damage but were convinced they could come up with a fix. One of the problems on the day they looked at the runway was the inches of snow from the previous two days and many of the runway cracks were filled with snow and frozen over, hiding the extent of the cracking. We also got a visit from the AAC civil engineering staff which had a few cold weather experts on the visitation team. They also were very concerned by the extent of the damage but one of the seasoned cold weather experts felt that as the temperatures warmed up the cracks would "heal" themselves a bit. They also recommended that we test fly the jets to see if the runway was structurally sound.

I was obviously very concerned about flying the jets to see if the runway was sound. I coordinated with the 21st TFW wing commander and got the OK to test the runway with the alert aircraft. Before the flight, I had a complete fifteen-member team of foreign object damage (FOD) walkers on the runway picking up any and all loose pieces of the runway they found. Finally, on February 7 we flew the jets, and they had no problem with the runway condition. The alert pilots flew a training mission and recovered at Elmendorf. The Elmendorf team could not get the jets turned in time due to a minor maintenance issue, so the jets remained overnight at Elmendorf. The next morning the jets returned, landed again with no problem with the runway, and were back on alert status.

Chapter Seven

Sex in the City. While I was not surprised that there could be some male/female issues while on a remote assignment, the following stories really challenged my leadership to address the underlying issues that were present during the entire year as the commander. The stories came from all areas of the base and involved military members in some cases and Air Force civilian employees in other cases; in a very unusual situation, one case involved a military member and a member of the Galena civilian community. If you remember, the first Commander's Gram that I got described every sexual situation that was taking place on Galena, but these following stories were not part of that original list.

In early October, two days after I took command of Galena, I got a call from the base operations folks that one of the new arrivals to the squadron was a very stunning looking young seventeen-year-old female airman. I was amazed that the Air Force Military Personnel Center would send a seventeen-year-old female to a remote Alaska base on her first assignment. My operations officer told me that I should brief this young lady on the "hazards" of Galena with nearly 300 lonely male airmen versus fifty female airmen. I asked her duty section to send her up to my office early the next morning. She arrived right on time and it was kind of awkward giving her a father-to-daughter briefing about the 300 lonely males that were at Galena. I told her she had every right to say no to anyone and advised her to be very careful in her relationships. Wow, my first father-to-daughter talk, and my own daughter at this time was only eight. Things seemed to be going well with the young airman, and I found that she was doing a good job in her duty section.

That all changed in mid-February when I got a call at 0330 hours in the morning from a very irate woman. The mother of this young airman called me in a very accusing voice asking me, "How could you let my daughter get pregnant?" Since it was early in the morning, I asked who she was talking about. She told me her daughter's name. I really didn't know that her daughter was pregnant. My answer to the woman was that it wasn't me that got her daughter pregnant, but I would look into the situation. The very next morning I had the young airman in my office asking her some very pointed questions. She immediately told me the father of the baby was one of our security police sergeants. Oh great, now I had to give a father-to-son talk. I called the security police unit and asked the guilty sergeant to be in my office in five minutes. He showed up and was confused why he was there. I laid out the information that I knew about the pregnancy. I knew that the male sergeant was married but when confronted with the situation he told me he wasn't married in his heart. I told him that was a BS answer and that I would contact the judge advocate general's (JAG) office at Elmendorf and explore the question charging him with adultery. I got the answer from the JAG that I didn't like much…adultery was a crime in the Uniform Code of Military Justice (UCMJ) but under current Air Force policy they would not prosecute any case of adultery unless it involved a minor. They stated that since the airman in question already turned eighteen there would be no judicial action recommended. I explained that she was seventeen when she got pregnant, but that held no sway with the legal folks.

I was really in a bind as I wanted kick the ass of the male sergeant but had no backing from the JAG. I also knew that the young female airman would not be able to finish her one-year tour at Galena as we had no capability to deal with a pregnancy in our small medical clinic. I called the sergeant back to my office and told him he was a jerk and hoped he told his wife of the deep shit he was in. I also called in the young airman and counseled her on her decision-making process and that this pregnancy would most likely end her future Air Force career.

The next situation was also something that I had never dealt with before. I got a call from the Galena Police Department that they were going to arrest one of my Air Force civilian employees. It seems this employee had beaten the crap out of his girlfriend in the visiting dormitory and was trying to leave Galena. I gave the Galena police the authority, in coordination with our security police, to come on base and arrest the civilian. I also authorized the local police to keep the civilian in their jail until bail could be worked out.

I needed to find out more about the situation and asked my housing officer and the civilian employee's supervisor to come to my office so we could get to the bottom of this difficult situation. The housing officer said that the civilian employee had a letter requesting that his "wife" be allowed to stay in his quarters with him. We had a policy that would allow a spouse of our military or civilian employees to stay on base with their spouse for no more than fifteen days during a calendar year. Obviously, the civilian lied on his application request. This situation was a different scenario for me, and I decided, after coordinating with the JAG office at Elmendorf, that we should let the local justice system take care of the assault charges and I would take action on falsifying the request document to let his "wife" visit the base. My decision was an easy one. I fired the civilian and banned him from entering the base. It took about three months to work our way through the civilian personnel system to get him fired but we finally did in early May.

Police versus Police. This next situation had me really scratching my head on what to do next. I got a request from the Galena chief of police to ban one his police officers from entering the base. Apparently, this police officer's wife was having an affair with one of my military people, and the chief of police was afraid that his police officer would try to either kill our military member or harass any potential witnesses in the case. My first step was to call in my security police chief and get his advice on what to do. I called the chief of police back and asked if he could share the military members name so we could do our own investigation into this situation. Once I got the military member's name, I called him into the office and made sure our security police chief was also there, and we asked a few very pointed questions. The military member was not married and was a sergeant in the supply unit. He said he had been having an intimate relationship with this civilian for about four months. I asked when he was to DEROS, and said he was scheduled to leave Galena in May. My security expert convinced this sergeant to stay away from the civilian female and just quietly slip out of Galena on his DEROS date. I called back to the chief of police and told him our plan. He thought it was a good one but wanted to make sure we enforced the ban on the base for his police officer so we could avoid a very critical situation. Fortunately, this plan worked, and our sergeant left Galena quite unnoticed. What really happened in this relationship was unknown, but nobody got killed in the process.

Minor Drinking Problem. In addition to the numerous sexual situations, the alcohol problems continued for our squadron. I attended a Galena city council meeting and was informed that the base was serving alcohol to minors from the city. That was quite a revelation. The first thing the next morning I called the club manager, the first sergeant, and the security police chief to my office to find out if this was a real problem or just another grievance from the city. Our long-time club manager believed that the situation described was caused by some of our young airmen bringing young female civilians from the city to the club to party. My security police chief asked if the bar tenders were checking IDs before serving drinks, which they were doing. However, it seemed the enterprising airmen would order the drinks using their legal ID and then take the drinks to the minors at their table. Obviously, this was not a good deal for our airmen or the minors. The first sergeant came up with a plan that he would observe the process in the club on the next Saturday night to see if that was the true scenario. Sure enough, the very next Saturday night he observed those very actions taking place. He got names of the airmen and also the names of the minors in the club.

The following Monday morning the three airmen were in my office at 0700 getting their butts chewed out for buying alcohol for the minors. I confirmed the names of each of the female minors that were illegally drinking, passed the names on to the Galena mayor, and banned those three females from entering the base. The security police had the list of the banned minors, and we never had another minor drinking problem again... at least none that I was aware of.

Poor Start to a Remote Tour. I got a call from the security police squadron commander at Elmendorf, a good friend of mine from Air Command and Staff College, that one of my inbound sergeants was arrested for driving while intoxicated (DWI) as he entered the front gate of Elmendorf. He had recently arrived in Alaska and was driving on base in a friend's car to set up transportation to Galena in the next few days. He blew a .25 BAT – well above the .08 DWI limit. The personnel folks at Elmendorf asked what I wanted to do with the individual. I immediately called the personnel office and told them under no circumstances were they to ship the individual to Galena. We had enough problems with the drinking issues at Galena already. The personnel folks told me they had no choice but to send the individual to Galena where we could immediately turn him around and send back to Elmendorf for judicial action. I raised the BS flag to our wing commander, explaining that I would not take the individual and that I also needed a replacement for him. Fortunately, the commander agreed and contacted the wing personnel folks and told them to stop the process of sending the drunk to Galena. It did take another three weeks, however, to get a suitable replacement to Galena.

On the River. I had done many reenlistment ceremonies over the years... including in the jungles of Vietnam, in the cockpit of a fighter aircraft, and during an official office visit. I had never done one as unique as I did in February 1989. A technical sergeant from the fire department stopped by my office and asked if I would consider doing to his reenlistment ceremony on the frozen Yukon River. I told him I would be glad to do it there if he was brave enough to walk on the frozen river for the ceremony. At the time, the Yukon River had been frozen for the last three months, and I knew after the record-breaking temperatures we experienced in January that the river was probably safe enough to venture out on to do the reenlistment ceremony.

So, on the morning of February 15 in balmy -40°F weather the reenlisting sergeant, our base chaplain, our first sergeant, and I started across the mighty Yukon River at a point that was about one mile wide. There were a few native ice fishermen on the river, and they told us the ice at the center of the river was approximately six to seven feet thick. No problem, I thought. It was very cold with a brisk breeze coming in from the southwest. I knew the wind chill factor must have been around -70°F. We all had our parkas on and once we got to what we thought was the center of the river I administered the reenlistment oath, although I had a hard time saying the oath because my lips and face were starting to freeze. The sergeant reenlisting felt the same freeze. We muddled through the oath, took many photos, and headed back to shore as quickly as we could. It was a memorable occasion for me, and I was thankful that all the remaining reenlistment ceremonies I did at Galena were on dry ground inside my very toasty warm office.

No Notice Sermon. I made a point of going to chapel service every Sunday I was at Galena. I certainly did not want to push my faith on anybody but I thought it was good leadership and to show good character by attending weekly services. I read scripture every once in a while and really enjoyed the sermons from our base chaplain. Since February was Black Heritage Month, we had a big banquet and many other events scheduled later in the week when I would be the keynote speaker for the banquet. Because I was only half awake during the Sunday morning service, I was surprised to hear our chaplain's aide say, "… and we are so pleased that Colonel Petitmermet has agreed to do the sermon this morning." I nearly crapped my pants as no one had coordinated anything with me. But here I was in front of eighty-five of my fellow squadron mates and I couldn't let them down.

Fortunately, I had some basic ideas, not a scripted product, in my head for the upcoming banquet later in the week. I got up and really had a very disjointed group of thoughts that I presented as the sermon. I think the only reason everyone did not get up and leave was because I was the commander. I couldn't get this sermon over fast enough and chew the butt of the person who failed to notify me that they wanted me to do the sermon. Apparently, the coordinator for the base Black Heritage month events forgot to communicate with the chapel and me, and the sermon request fell through the cracks. I was continually bugging the personnel sergeant who forgot to tell me that I was to do the sermon for the rest of my time at Galena. The good news was that my speech at the Black Heritage Month dinner was well received and we had about one hundred folks attend with a very good mix of black and white service members. My talk had a very patriotic theme rather than a black and white exposé. The dinner speech was much better than the one and only church sermon I have ever given, and I don't intend to give any more.

More Racial Issues to Deal With. Unfortunately, some very intense racial issues re-surfaced from the same senior master sergeant in the supply department. My Commander's Gram Box had seven very detailed complaints about the actions of this senior enlisted person. Everything from daily duty assignments within the supply shop to some very blatant racial comments, to others within the squadron. The pot boiled over when I received five end-of-tour (EOT) out-briefs about more bad behavior from the same individual. I was forced again to bring her into the office, counsel her on her behavior, and explain that senior enlisted people are supposed to be part of the solution not part of the problem. She complained once again that there was no senior enlisted advisory council in the squadron and that we should start one immediately. I had tried that approach in October the last time I counseled her and she stated then that she was too busy to be on the advisory council. So this time I ordered her to join the council or keep her mouth shut and start behaving like a senior NCO. She reluctantly agreed but the feedback I got from the other members of the advisory council was that she just sat and glared during the last meeting. The situation would escalate again in May.

My 42nd Birthday. Just like all the other folks at Galena, being away from my family during important events in my life was a bummer. I celebrated my 42nd birthday that year at Galena, alone and away from my family. I guess I was at least with my squadron family but that is just not the same. I had scheduled my normal weekly officer's staff meeting for the lunch hour on my birthday, and I was well into my agenda when there was a bit of commotion at the door to the commander's dining room. I looked up and to my surprise there was my lovely wife, PK.

PK had surreptitiously scheduled a commercial flight to Galena and coordinated with my female administration clerk to get picked up at the commercial terminal and remain hidden on base until my staff meeting. It was the surprise of my life and a welcomed "make my day" event. When she walked into the lunch staff meeting, all I could say was, "Holy shit, what are you doing here?" We had a great afternoon, and I walked her around explaining all of the different functions on the base.

We had dinner at the club, and the staff set up a quiet booth with candlelight and champagne for us. The club staff really out did themselves – the food was perfect and the atmosphere superb. Of course, I was called on the radio for some emergency and had to leave for a very short time. My wife waited for me at the bar while I checked on the emergency. In a humorous turn of events, while I was gone from the club one of my civilian employees thought my wife was there by herself and tried to pick her up. What a funny response from him when he found out she was the commander's wife. As soon as I returned to the club most of the squadron members who were not working surprised me with a birthday cake in the ballroom. What a great day to be alive and the commander of such a great squadron.

The very next night I was asked to be one of the judges for the local school talent show in the city of Galena. I took my wife with me, and we had a very fun time. That evening, the local Alaska wildlife officer asked my wife and me to stop by his home after the talent show because he wanted to show us a unique sight at his house. The night was a very cold one, around -40ºF. I was very curious as to what the wildlife officer wanted to show us. He took us to the back of his property where his horse was stabled. He had been feeding the horse some very high fiber grain for the cold weather. He was so proud of his horse and the way the horse farted! I asked him, "Is this what you brought us out to see?" Just then the horse let out a very loud and long fart. It was so cold that the horse created about a fifteen-foot contrail right out of his butt. We all had a very hearty laugh.

The very next morning my admin clerk told me that I had a call from the National Inquirer Magazine to talk to me about the cold weather and how we were coping. I was so tempted to tell him about the horse fart but held my tongue and told him some BS about what we were doing to combat the cold.

All too soon, my wife had to fly back to Elmendorf, but what a very memorable birthday she brought me. The squadron fire department gave her the Firedog send-off with water spray and all the sirens blaring. It was a spectacular departure for my wife and everyone else on the commercial flight.

February Intercept Summary
    February 27: 1 IL-20 Coot M

Chapter Eight
"NORMAL OPERATIONS"
March 1989

More Personnel Issues. The leadership challenges continued with some events I had never experienced as a leader in the Air Force. First, a young sergeant arrived at Galena with an arrest warrant for stealing government equipment from his previous base. My initial question was why did the previous unit release him for assignment? A quick consultation with the security police and the JAG at Elmendorf told me we needed to take some action on this individual. "Duh!" was my answer. We immediately "arrested" him again at the barracks and took him into custody. My security police were then directed to escort him back to Elmendorf where he would be incarcerated and processed for his theft crime. Galena had neither the capability to incarcerate him nor the ability to provide him the legal advice he would definitely need. We got him out of Galena on the very next commercial flight the very afternoon he arrived. I believe this was the shortest stay ever at Galena for a person assigned to my squadron. I was immediately on the phone to our personnel chief at Elmendorf asking him to put in place a better screening process for the folks being assigned to our remote bases.

Iditarod Sled Dog Race. One of the primary events in Alaska each year is the 1,100-mile sled dog race from Anchorage to Nome, Alaska. There are many check points along the incredibly difficult course, and in 1989 one of the check points/rest stops was Galena. The check points are manned mostly by volunteers and our squadron was able to provide a very limited supply of volunteers due to our preparation for an upcoming unit effectiveness inspection (UEI). Our volunteer tasks involved helping move the straw bales for the dogs to sleep on and pre-positioning the food for the dogs during their rest stop. Due to my busy schedule, I was only able to visit the rest stop one time during the race.

The Iditarod Trail Sled Dog Race has been held on the first Saturday of March every year since 1973. The race route stretches from Anchorage to the western coast of the Bering Sea at Nome and covers a length of around 1,100 miles. Men and women from all walks of life compete in this grueling race as mushers. Mushing is the act of transporting people or goods with the help of dogs, and the driver of the dog sled is called a musher. The race involves the musher and his or her team of dogs competing not only against other racers, but also against the harsh forces of nature in the tough Alaskan terrain.

In 1925, a diphtheria epidemic in Nome required the immediate delivery of diphtheria antitoxin. Since the serum couldn't be flown in, the twenty-pound cylinder containing the serum was transported via train from Seward to Nenana. Twenty mushers along with more than 150 dogs relayed the serum from Nenana to Nome across a distance of 674 miles. Five and a half days later, a Norwegian musher arrived at Nome with the serum, helping to stop the spread of the diphtheria epidemic. The annual Iditarod is run each year to honor the brave mushers that delivered the serum to Nome in 1925.

Near Death Accident. A civilian contractor working on an electrical project on the base was accidentally hit in the head by a piece of the equipment being used to install the new electrical components. My medical tech and I were called to the scene to start the process of getting the civilian evacuated off base. It looked really bad as the individual was convulsing and not conscious. My med tech said we needed to get the individual back to Fairbanks or Anchorage as soon as possible. As I was doing the coordination for the airlift out of Galena, I got a call from the wing commander asking me about the electrocution fatality that had just occurred. My response was that there was an accident but the civilian was in the process of being evacuated and it was not an electrocution but a head injury. I was very curious how the wing heard there had been a fatal electrocution at Galena.

I was able to request a C-12 aircraft at King Salmon to be diverted to fly into Galena and pick up the injured civilian. My next call was to the command post at Elmendorf trying to determine how the wrong information about the accident was sent to them. The command post told me that my disaster preparedness chief had called the command post and reported the electrocution. Wow, did I have some work to do with my staff. I called an emergency staff meeting with my officers and senior enlisted folks and really read them the "riot act" about reporting things from Galena that were not correct. From that point on, I would approve any and all reports going back to the wing at Elmendorf. As for the accident, it was a freak event caused by some inattention, some ice on the ground, and a disregard for obvious safety factors. I asked my ground safety NCO to get more data on the accident so we could brief the circumstances to the entire squadron at the next commander's call.

Chilly Easter Service. Our chaplain requested that we conduct a sunrise service on Easter Sunday, March 2. The chaplain had set up for the service on the base softball filed and had about fifty chairs, a sound system, and all the trimmings of a makeshift altar. The morning was clear and -25ºF with about a -45ºF wind chill and strong north wind. There was only eight of us at the service including the chaplain. We all were wearing our winter parkas and arctic gloves – and shivering our asses off. The cold weather "froze" the portable music box that was trying to play Easter music and thankfully the chaplain gave a very abbreviated sermon. We all huddled around each other trying to stay warm and then adjourned to the dining hall for more fellowship and sharing about our past Easter celebrations that were part of our family heritage. It was the coldest Easter celebration I had ever experienced but still a spiritual event.

Later that Easter Sunday I worked at the dining hall grilling, cooking, and serving nearly 300 steaks to our squadron. It was one if the pleasures of being a commander and being with the folks on important holidays. Three other senior enlisted members also helped with the steak grilling, and it was another huge success.

Family Visit. My brother, a former Air Force officer, paid me a surprise visit while he was in Anchorage and Fairbanks on business. He just scheduled a commercial flight into Galena to see what his little brother was doing. What a very pleasant surprise. I cleared most of my schedule and gave him a complete behind the scenes tour of Galena, with lunch at the dining hall, dinner at the club, and then a movie at the base theater. The move was Bat-21, a story about a Forward Air Controller (FAC) during the Vietnam War. I had been a FAC during the Vietnam War and was able to fill in some of the Hollywood's omissions and distortions from reality. I was able to get my brother a night in the VIP suite in the officer's barracks. He was pleasantly surprised by the first-class accommodations at our remote site. We stayed up and talked until nearly two the next morning. I just hoped that my three children would have the same relationship that I enjoyed with my brother.

The next morning, we ate at the dining hall and then did a tour of the entire base, the old deactivated Campion Radar Station about twelve miles east of Galena, and then the downtown area of Galena. Since my brother was in the retail sales business, he was very interested in the local store. He was just amazed at the prices of most items available. But I told him that was not unusual as all the supplies had to be flown into Galena because there are no roads from civilization to us. After the visit to the store we walked halfway across the mile-wide Yukon river that was still completely frozen. I believe I gave him the true arctic experience during his short stay. He got on the afternoon flight out of Galena that same day but what a pleasant surprise his visit was.

Supervisor of Flying. One of the unique tasks at a remote site for the squadron operations officer was to act as the supervisor of flying (SOF) whenever the alert fighters were airborne. With only two aeronautically rated officers (pilot or navigator) on station, the operations officer, or me as the commander, had to do that duty. Under normal conditions that duty was exclusively the task of the operations officer. The primary function of the SOF was to serve as the wing commander's representative whenever flying activities took place and to be responsible for the safe and efficient mission accomplishment of the unit. The SOF had the authority to cancel flying activities, divert aircraft, dictate mission requirements, approve specific activities, coordinate for airport or airspace utilization with FAA, and direct emergency actions involving wing airplanes. Additional tasks included monitoring weather conditions, airfield status, barrier configurations, runway condition readings (RCR) status, and any other factors that could affect a safe recovery of the alert aircraft. The SOF would also act as the primary coordinator for any in-flight emergency support the airborne pilots may need. It is a very vital function and is in use at all Air Force flying locations.

The 21st Tactical Fighter Wing had an operating instruction (OI) that really spelled out the exact duties of the SOF. Those items are summarized below:

1. Be responsible to the Wing Director of Operations (DO) for monitoring and supervising all phases of unit flying operations and provide guidance, advice, assistance and recommendations to aircrews, unit supervisors and/or other supporting agencies regarding the safe and efficient conduct of flight operations.

2. Ensure that in-flight emergency (IFE) recovery plans and weather-related mission changes reflect sound airmanship, follow Air Force Instructions and technical order (TO) guidance, and adhere to sound Operational Risk Management (ORM) principles.

3. Direct appropriate actions, on behalf of the Wing DO, to correct/prevent unsafe situations. This includes the use of any and all resources to include radios, FM nets, telephone hot lines and all wing-flying operations on the ground or in the air.

4. Prior to the first launch, ensure the airfield status is suitable for safe operations IAW Air Force, major command (MAJCOM) and local directives.

5. During an emergency or an abnormal situation, provide aircrews with guidance, timely advice and assistance to determine a correct course of action.

6. Be in a position to visually monitor the final approach and landing of In-Flight Emergency (IFE) aircraft.

7. Monitor the status of primary and emergency airfields and inform aircrews of changes that may affect flight operations.

8. When deteriorating weather conditions affect flying operations (which they did quite often in Alaska), coordinate with wing agencies and utilize all available resources (i.e., weather, radar, tower personnel, pilot reports, etc.) to determine the best course of action for wing aircraft.

9. The SOF will determine suitable weather alternates and inform the Wing DO accordingly.

10. Coordinate with the air traffic control (ATC) watch supervisor or senior controller for runway changes as needed.

11. Prepare a daily log to aid in tracking operations and major events.

12. The SOF will debrief the Wing DO of any aircraft involved in an unusual situation, IFE, weather divert or other mission change requiring SOF action or intervention.

13. Be the primary liaison between Ops and Maintenance during the execution of the flying schedule.

14. Ensure flight crews are briefed on the following: Airfield status and configuration, scheduled and available airspace, applicable weather for locations that aircrew are flying, significant local hazards.

15. The primary objective during an abnormal/emergency situation is the safe recovery of the aircrew and aircraft.

During March of 1989 I had to assume the SOF role for a total of two weeks as my operations officer was on temporary duty (TDY) at a base in Florida conducting a death investigation of an F-15 pilot. Being the SOF really impacted my duties as a commander but it was a task that had to be done. I had my admin staff bring all the commander's issues to me at the combat alert cell (CAC) because I could do both tasks as long as I was not dealing with an aircraft emergency, deteriorating weather, or airfield support equipment failures. I was at least comfortable doing the SOF duty as I had done it many times at Elmendorf when I was a flying squadron commander. Fortunately, no significant emergencies came up during my SOF duties other than a couple of the fighters diverting back to Elmendorf with various mechanical issues.

Wing Non-Support. Another prime example of my "leadership by wandering around" concept uncovered a real problem within the squadron. Certainly it was not a life changing problem, but it was one that nevertheless affected the morale of the entire squadron. One support area that we did not have at Galena was a laundry and dry-cleaning service. Each member of the squadron, including me, was expected to launder their own clothing by using washers and dryers that were located on each floor of every dormitory. The expectation was that only residents living on that floor would use those particular washers and dryers. As I made my rounds through the dormitory one evening, I was approached with the problem of many non-operational washers and dryers. I called my first sergeant and together we did an inventory of all the washers and dryers in all the dormitories, including the officer's dormitory building, and found that out of the eighteen washers and dryers on the base only three washing machines and four dryers were operational. That certainly was not a good situation for morale or for maintaining a polished military appearance.

I contacted the supply officer early the next morning and told him to get this problem fixed immediately. The 21st TFW supply division had a dedicated senior civilian employee that was responsible for all supply support for the forward operating bases and remote radar sites. I told my supply officer to call him to get the necessary washing machines and dryers purchased and put on the next available aircraft flying into Galena. In about an hour my supply officer informed me that the 21st TFW supply coordinator for the forward operating bases would not provide any additional washers or dryers and that the squadron personnel could just take their laundry to the laundromat in Galena City. I was outraged at the response and immediately contacted the wing commander describing the problem at Galena. He told me to work it out with the supply folks, so I called the deputy commander for resources at Elmendorf, a colonel, and described the problem and the non-support answer from his supply expert.

It seems as though the deputy commander for resources had heard many complaints about site support in the last six months, and he said he would send the site supply coordinator, a senior government service employee, to Galena on a C-12 flight the next day. Meeting the supply coordinator as soon as the aircraft landed, I took him to the dorms and showed him all the broken down washers and dryers, then took him to Galena City to show him how using the city laundromat, with their three operational washers and one operational dyer, could not support our 350 military and civilian members of the squadron. As it turned out, this great site coordinator had not visited either of the forward operating sites for the past three years and had no clue about what was and was not available at the sites. His plan for our airmen to take their laundry to the city was absurd as there were no privately-owned vehicles on the airbase. How could he expect the airmen to get their laundry the four miles to town in -70°F weather? I told him I expected the washers and dryers to arrive within the next twenty-four hours or I would file a complaint with the 21st TFW inspector general over his nonsupport of the sites. The very next day ten new washers and ten new dryers arrived on the normal resupply aircraft, and we were back in business. This was certainly not the last time I would have problems with supply support at Galena.

March Intercept Summary
    March 14: 2 Tu-95 Bear Gs
    March 21: 2 Tu-95 Bear Hs

Galena in the Winter Overhead

Birchwood Hangar

Tom Petitmermet

Combat Alert Cell

F-15 Armed for a Mission

F-15 Taxiing in the Snow

Tom's Wife Visits Galena

FPS-117 Radar

Galena Headquarters Building

Cool Barge Delivery

Typical Winter Day in Galena

Randy Travis

Patty Loveless

Doing Our Mission

Bear Hs and F-15

Bear H

Soviet Coot

C-5 Landing at Galena

17 January Thermometer

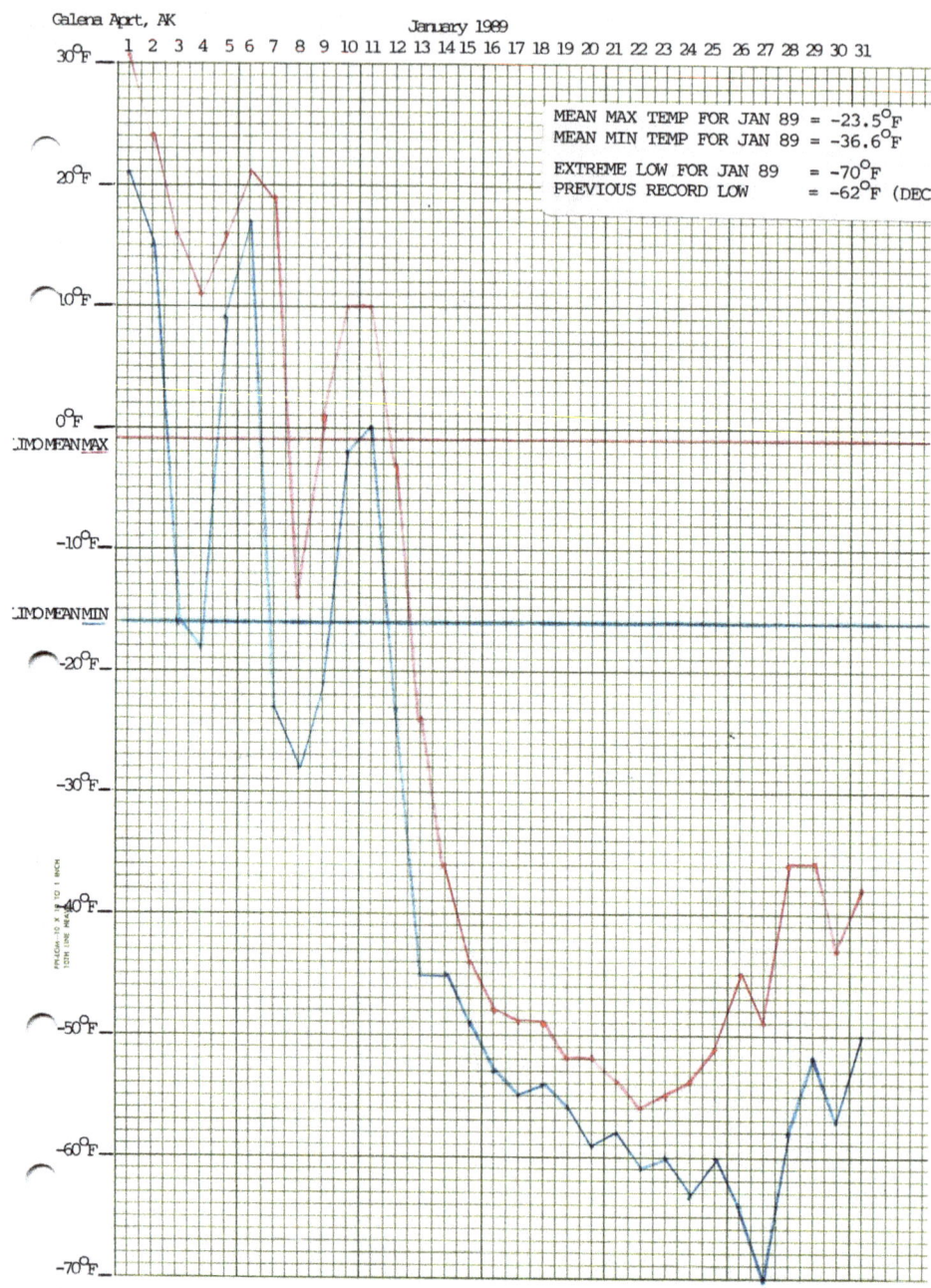

Deep Freeze, January 1989

```
GAL         SA       1855Z
            RPT
  SKY:    -X  1  SCT
  VSBY:   10
  TEMP:   -70F     DWPT:M__F
  WIND:     CALM          KTS

ALSTG:    30.08INS

  RMKS:   F3 VSBY W-N5/16
          TWR VSBY 1    ECT  -70
    PA:   0
07/RC_
```

```
GAL         SA       1155Z

  SKY:    -X
  VSBY:   3IF
  TEMP:   -53F     DWPT:M   F
  WIND:   260/05          KTS

ALSTG:    29.77INS

  RMKS:   F1 VSBY N1    ECT  -88
    PA:   +290
```

Weather Observations, 27 January 1989

Commander's Vehicle

Cracked Runway

Runway Cracks II

Eskimo Scouts

Iditarod Dog Sled Race

Flood Gage Low Reading

Flood Gage High Reading

Ice Chunks Near the River

Vehicles on the Dike

Flooding at the Fuel Tanks

Widespread Flooding

Ice in the River

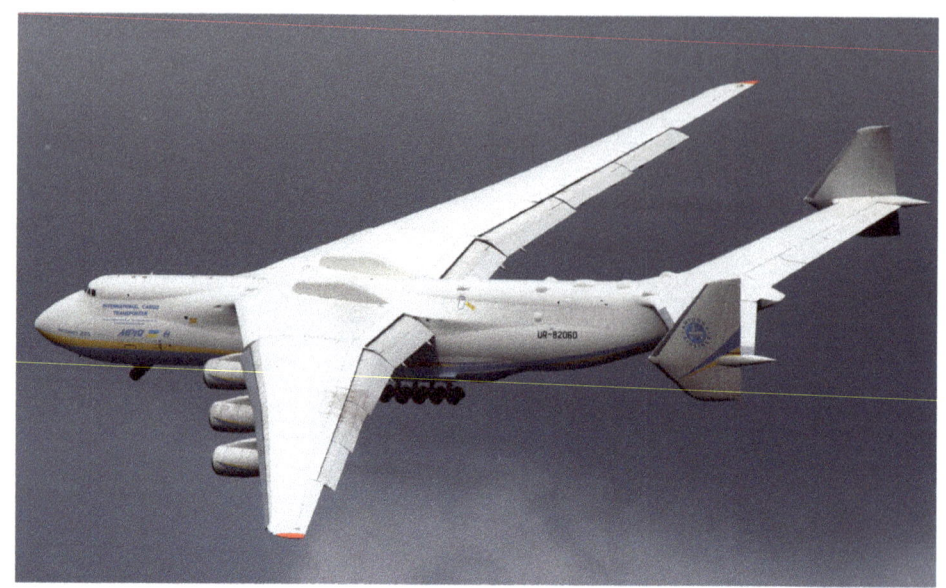

Soviet An-225

Soviet Mig-29

Fishing Boat

Tom's Catch of the Day

Chapter Nine
THE THAW BEGINS
April 1989

"**N**ormal" Operations. Following another cold month in March we were finally getting our day-to-day operations back to normal. The threat of a unit effectiveness inspection (UEI) was still very much on the horizon. The Alaskan Air Command inspector general team had proposed an annual squadron inspection schedule of April 8-16. I was certainly none too pleased with those dates as we were still in the midst of cleaning up and repairing all the damage we had sustained from the arctic freeze in January and early February. The UEI is a commander's report card on how well he is leading the squadron and completing the assigned mission. I knew our squadron was doing well but wanted to make sure we had an even-handed attempt to show our true skills. I appealed to the wing commander to delay the UEI until May so we could address the many cold weather issues we were dealing with. Miraculously, we did get a reprieve on the UEI date until May 15-23, 1989.

Butt Patrol. As the weather improved and we started to see temps occasionally above freezing, the thaw began. One unique aspect of the Alaska thaw was we started to see some of the trash that was thrown in the snow throughout the snow season. Many times during the snow season I saw individuals who were smoking just throw their cigarette butts in the snow. Things began to become unsightly. Working up a plan for the base cleanup, I asked my officers and first sergeant to be very observant over the next couple of weeks and identify, by name, any individual who they saw throw their smoking material in the snow. These individuals would become the main clean up cadre, the Butt Patrol, when we did the base spring cleanup. After the two-week observation period we had a very robust team who would be on the butt patrol to help with the base clean up.

Not My Job. Unfortunately, the fire chief at Galena was just not providing the leadership we needed for that vital function. I had thought many times during my months at Galen that he needed to be replaced. He stopped by my office one day and said he needed to talk to me about a very urgent issue. I could hardly wait to hear what that issue may be. Quoting from Air Force regulations, he stated that members of the fire department did not have to do any additional duties unless they volunteered to do them. Fortunately, my first sergeant had heard rumors that the fire chief would bring this very complaint to me so I had my response prepared. You see, a number of his fire department personnel had been selected to be on the butt patrol team.

One of the unique situations at our remote base was that our military members had the opportunity to take off-duty jobs at the club, the bowling alley, the Morale, Welfare and Recreation (MWR) unit, or a few others locations. The interesting aspect of this program was that the unit commander, in this case me, had to approve each off-duty work request. I had reviewed the Air Force regulation and found that not only did I have the authority to approve each request, I also had the authority to rescind already approved requests. I asked the fire chief if he really wanted me to follow the additional duty regulation for the fire department personnel. He said absolutely and that I must immediately take all his people off the butt patrol roster. I told him OK…done deal.

Before he left my office, however, I informed him of the off-duty work regulation and the role I had in approving all off-duty work requests. He looked rather confused. He himself was working off-duty at the club as a bartender. I told him that effective immediately I would implement the no additional duty assignments by the letter of the regulation for the fire department and at the same time would rescind and disapprove all current and future off duty work requests from the fire department. His answer to me was, "You can't do this." I told him, "I just did!"

Boy did the excrement hit the fan! It wasn't thirty minutes later when my phone was ringing off the hook with complaints from the fire department personnel about my "new" off-duty work request policy. I thought for about an hour that I was going to have a mutiny on my hands. The new policy only applied to the fire department as they had the regulation stating they could only do additional duties if they volunteered. The remaining members in the squadron had no such regulation, and my new policy did not apply to them. I had them by their fire hoses! It was but one hour later that the fire chief was in my office saying his troops, including himself, would "volunteer" to do any additional duty requested if I would not block or rescind any fire department personnel off duty work requests. I believe they got the message.

One of the key leadership lessons I learned from this exercise was to really listen to the senior enlisted leaders in the unit. They had the direct pulse of the squadron, usually knew all the rumors, and always came up with the best solutions. I carried this lesson with me for the rest of my Air Force career.

The Thaw Begins. The weather finally began to warm a bit although we did have just a few freezing nights and mid-30°F days. The piled-up snow began to thaw, and the runoff began to migrate to the low areas of the base. We had numerous buildings that started to get melting snow runoff entering the buildings. I met with my CE officer, and we came up with a plan to monitor the melting snow so we would preclude any further damage to our buildings beyond what was caused by the arctic freeze.

An important corollary to the thaw was making sure the squadron was ready to evacuate if the Yukon River flooded Galena during the ice breakup. During the review of existing procedures, we found some old 1982 checklists in place that I wanted to review to make sure our preparations were completed in a timely manner if a flood did occur. The breakup of the Yukon River is an annual event and can have catastrophic consequences as the massive amount of river ice starts its normal flow to the Arctic Ocean. In many years the ice is so thick that it backs up in choke points along the way to the ocean causing massive flooding behind the ice jams.

I directed that we start a daily, rather than weekly, commander's staff meetings and asked each officer to review and practice their part of the functional checklist to prepare for a potential flooding situation. I wanted each function to really test out their part of the response checklist.

As we progressed with the checklist review, we found numerous deficiencies in the process and attempted to correct and test the new procedures as we went. One of the most difficult tasks was to account with 100 percent accuracy the number of personnel assigned and present at Galena on any given day. That number would be critical as we determined who would stay behind to be a caretaker of the base and who would be evacuated to Elmendorf. Every morning we seemed to have a different number present at Galena. That did not make me happy. I knew the personnel folks were trying, but I didn't think they were getting the help they needed from each functional area.

The next critical step was to determine the correct balance of staff that needed to stay behind if we did flood. That number fluctuated almost every day as the functional areas fine-tuned their needs and determined exactly the type and number of experts that would be needed to accomplish their critical base function if they remained after the flood. The final number of critical personnel that would remain at Galena during an evacuation was fifty-one.

My first sergeant told me that the airmen thought this exercise planning for a potential flood evacuation was a bunch of BS and that we would never have to evacuate the base. I thought this piece of the planning process was critical as each individual who would be evacuated would have to bring the things they needed to survive at Elmendorf for up to two weeks. Again, my wise first sergeant had a perfect plan. Just before the end of April the first sergeant made up a checklist of all the things each person should bring with them if we got evacuated due to flooding. The list was passed to each functional area leader who were told to distribute it to every service member on Galena. After about a week, we did a practice recall for a mock evacuation and every member who was to evacuate had to bring their personal belongings and report to base operations to be processed for the flight to Elmendorf.

In addition to the personnel processing, we also practiced the evacuation checklists and transported the critical equipment that was designated to be moved to the top of the dike. We did find some parking and dike stabilization issues when we moved all the critical equipment into place on the dike. We reevaluated the vehicles and equipment that we thought we needed during the flood and reduced the total pieces of equipment and vehicles by about one third so they could all fit-on top of the dike out of any potential flooding areas. We were ready to implement the evacuation plan if necessary.

Song Request. On the early morning of April 15 I got a call from the senior alert pilot about an incident that happened the night before at the CAC. I immediately went out to the CAC to find out what the problem was. In the early morning, 0330 hours, someone was banging very loudly on the exterior door leading from the ground level to the second floor of the CAC. This door is just below the pilots sleeping quarters. Both pilots were awakened by the banging on the door and were concerned about their required crew rest. I went into the CAC cab and asked if the controllers on duty knew anything about the banging on the door earlier in the morning. The controller said he knew about the incident and that a security police member was trying to get the CAC operators attention to call the Alaska radio network and request a specific song for the security policeman's girlfriend. That was a bad choice for the security policeman. I called the security police superintendent and ordered him and the SP culprit to my office immediately. The SP arrived, and I asked to get his side of the story. He told me he just wanted to dedicate a song to his girlfriend and that at no time did he leave his post and the observation position of the alert fighters. I told him that I would do some more investigation as his action caused us to put the fighters on a mandatory scramble order (MSO) status because he had interrupted the pilots' crew rest.

I then went back out to the CAC and tried the door he was banging on to see if he could still observe the fighters from that position. The answer was very obvious…he left his post uncovered and he could not see the fighter jets from the door he was banging on. I really had no choice but to punish him to the maximum allowed by the UCMJ. Coordinating with the JAG back at Elmendorf, we came up with a plan to reduce him by one rank and fine him $250 for two months. I completed the paperwork and then it was time to call him back to my office to administer his punishment.

He reported right on time, and I had my first sergeant in the room with me when I advised him of his punishment for leaving his post at the CAC. I told him that I personally reviewed his claim that he could still see the fighters from the back door to the CAC. I told him that there was no way to see the fighters and that I was going to reduce him by one grade and give him a two-month fine. He was standing at attention in front of my desk the entire time. I then told him that if our defense condition (DEFCON) had been two levels higher I could have him executed for leaving his post under increased DEFCON status. He immediately became very rigid and fainted right to the floor. My first sergeant caught him just before he hit his head on the coffee table in front of my desk. I guess I really got his attention. Once he came to, I told him to just get out there and do the best job he could as he still had nine months left on his one-year tour.

Sneaking onto Galena. It is highly unusual for someone in the active duty military to sneak onto a base for an assignment. I had been notified that a staff sergeant had arrived at Elmendorf for assignment to Galena but had an outstanding warrant for his arrest for a crime (felony) committed at his previous base. He had travelled to Alaska on forged travel orders and was trying to in-process with the Elmendorf personnel office to transfer to Galena. I told the Elmendorf personnel folks that under no circumstances would I accept the staff sergeant in my squadron. Early on the morning of April 21 I got a call from my personnel office that the suspect staff sergeant had arrived at Galena and was trying to in-process to the squadron. How could that have happened? I called my chief of security police and we both headed over to the personnel office in to investigate this unusual situation. It seems that the staff sergeant in question had forged some additional travel orders with the hope of arriving at Galena undetected, hiding in the squadron, and avoiding the warrant for his arrest. Not so fast. I immediately contacted the personnel office and the security police office at Elmendorf and told them I had one "prisoner" en route back to Elmendorf for incarceration and confinement. I tasked one of my security police sergeants to escort the sergeant back to Elmendorf and not let him out of his sight until he was securely in custody. Needless to say, I had some very strong words for the personnel folks back at Elmendorf about this situation.

Fire at the Base Gymnasium. I got a call in my quarters at 0245 hours that the base gymnasium was on fire. When I arrived at the gym, I saw flames and smoke pouring very heavily from the side of the building. The fire and smoke appeared to be coming out of the side of the building where the main electrical supply lines were located. The fire department was already on scene and decided to cut the power to the building in coordination with my civil engineering folks. As soon as the power was cut the flames and smoke subsided a bit. However, the fire continued until the fire department personnel removed all the insulation around the electrical opening in the building. The fire was finally extinguished but we still had work to do. The gymnasium also housed the bowling alley and a snack bar. We coordinated with the fire department, the MWR folks, and the supply folks to remove all of the food from the snack bar since the power would be out for a few days. I finally got back to bed about 0500 but was awaked at 0630 by a call from the command post at Elmendorf from the wing commander asking about the fire and what corrective actions we took. The civil engineering folks isolated the problem and did the necessary repairs, and the gym was back in operation in less than three days.

"Sick" and Missing Leadership. As we were in the midst of preparing for the upcoming unit effectiveness inspection (UEI) I needed all hands-on deck to ensure all the preparations were taking place in a timely manner. Just as things were starting to come together, I was faced with four critical medical emergencies. First, I was notified that my operations officer had severely sprained his ankle playing basketball and needed to be transported back to Elmendorf for treatment. We were not sure if he broke the ankle, but I knew we could not treat him at Galena. Once he got to Elmendorf, I was advised that he might be out of service for six weeks and might not be able to finish his one-year tour at Galena. As I mentioned in a previous chapter, that meant that whenever the alert aircraft flew, I had to pull SOF duty – I would be away from my office, not leading preparations for the inspection.

The very next day I was notified that my vice commander was very sick and had reported to the Galena clinic for treatment for a bowel blockage. This blockage was caused by a previous gunshot wound that he had suffered in 1974. This was certainly not a condition that we could treat at Galena. We coordinated transport to get him back to Elmendorf for treatment. I contacted his wife in Washington, D.C. to let her know that we were taking good care of her husband. Reports from the hospital at Elmendorf indicated that he might need surgery and could be out of service for one month or more. Wow, both of my key officers, both majors, were now out of service for at least the next month or more. Next, my senior captain, the chief of maintenance, was notified by the Red Cross that his grandfather had died expectantly and he needed to depart Galena for the funeral. Just as I got my vice commander on the aircraft for evacuation, I was notified that my assistant civil engineering officer, a first lieutenant, had broken his foot after dropping a fork lift blade on it. Finally, the master sergeant in charge of transportation fell off of his motorcycle and broke his collar bone putting him out of service for two weeks. Preparing for the two upcoming events – a potential base evacuation and the UEI – without four key leaders brought on new meaning, and I was hoping some of the young airmen would step forward and take some leadership responsibilities. Fortunately, I did have a lot of previous experience in the operations area and also had tour as an aircraft maintenance officer in a F-106 squadron, so I knew the basics of those two functions. Whether I could lead those two areas (operations and maintenance) and function effectively as a commander was still to be seen.

C-5 Arrival. I was notified that we would be receiving a C-5 aircraft delivering supplies to Galena on April 15. A C-5 had never landed at Galena before, so this was a big deal. A C-5 is the largest transport aircraft in the U.S. Air Force inventory and is a massive airplane. It has a wingspan of more than 222 feet, a length of more than 247 feet, and a height of sixty-five feet. It had the capability to carry 270,000 pounds of cargo and had a max takeoff weight of 840,000 pounds. It is a huge aircraft. Since my operations officer was not on station, I coordinated with the airfield experts back at Elmendorf to confirm that the

Galena runway, after the cold weather incidents we had in January, could hold the weight of the C-5. I was also concerned by the weight bearing capacity of the parking ramp on the airfield. I got a "good to go" recommendation from the airfield engineers at Elmendorf for the C-5 to land at Galena. I had all my enlisted operations personnel and flight line maintenance personnel on the ramp and runway the morning of the C-5 arrival making sure there were no obstacles for the aircraft to land, taxi to the ramp, and off-load the cargo. I also confirmed that we had the capability and the fuel necessary to refuel the C-5 after landing.

Right on schedule the C-5 made its approach and safe landing at Galena. Word had quickly spread around, not only on the base, but also in town as there were probably seventy-five civilians at the commercial air terminal to see the arrival of the C-5. After the aircraft arrived and parked, I met the crew of the C-5 and welcomed them to Galena. The C-5 aircraft commander told me that they had about an hour and a half before the scheduled departure, and they would be happy to open the aircraft up to any of the observers that wanted to see the inside. You would have thought it was Christmas as I sent my first sergeant over to the civilian terminal and he told the observers that they could look inside the aircraft if they wanted to. Nearly all of the civilians and most of my military folks that were watching the arrival of the C-5 were able to tour of the aircraft. Right on schedule, the C-5 taxied out to the runway and had a very normal takeoff and departure.

Supervisor of Flying (SOF) Again. While my operations officer was back at Elmendorf having his badly sprained ankle treated, I again had to pull duty as the SOF. On April 13 and April 15 we had two intercepts just two days apart. The first occurred on April 13 when the two Galena alert F-15s were scrambled to intercept a suspect Soviet aircraft northwest of Point Barrow, Alaska. Our long-range radar system detected a single aircraft approaching the ADIZ while it was still in international airspace. A KC-135 tanker was also scrambled out of Eielson Air Force Base in Fairbanks, Alaska to support this mission and was in position to refuel the F-15s after they completed the intercept and identification of the Soviet aircraft. The F-15s identified the aircraft as an IL-20 Coot. The Coot is the North Atlantic Treaty Organization (NATO) designation for this four-engine intelligence gathering aircraft. The aircraft is 117 feet long with a wingspan of 122 feet. It can fly at 410 knots and has a side-looking radar surveillance system on board.

The intercept on April 15 was another Coot aircraft which was intercepted west of Point Hope, Alaska. Like the intercept on the 13th, the Soviet aircraft never entered the ADIZ and turned away from the coast after being intercepted. This Coot aircraft was an M model that was used for scientific research by the Soviets. The F-15s were again supported by the KC-135 out of Eielson Air Force Base and an airborne warning and control systems (AWACS) aircraft flown out of Elmendorf Air Force Base.

The AWACS E-3 Sentry is a modified Boeing 707 airborne warning and control system with an integrated command and control battle management feature providing surveillance, target detection, and tracking. The aircraft provides situational awareness of friendly, neutral, and hostile activity. The large rotating radar dome on the top of the aircraft provides radar coverage and control to a range of more than 250 miles.

April Intercept Summary
    April 6: 2 Tu-95 Bear Hs
    April 9: 2 Tu-95 Bear Hs
    April 13: 1 IL-20 Coot
    April 15: 1 IL-20 Coot M

Chapter Ten
DISASTER ON THE RIVER
May 1989

Assistant Secretary of the Air Force Visit. I was notified that the Assistant Secretary of the Air Force for Personnel would be making a visit to Galena on May 2. It was not a big deal as we had numerous dignitaries visit Galena during my time as commander. We had all the necessary preparations in place and had fine-tuned our squadron briefing. All was set for a fairly routine visit. The briefing that I prepared went well, I thought. The visiting dignitary seemed quite bored with the whole visit. When I asked if she had any questions her only response was how many females were assigned to the squadron and that she would like to visit our day care center on base. I guess she missed the part of the briefing that described the personnel in our squadron as unaccompanied active duty members serving a one-year remote assignment. An unaccompanied tour meant that you could not bring your dependents to your assignment. That concept went completely over her head.

I next gave her an in-depth tour of all the facilities on the base and tried to introduce her to as many females as I could during the tour. At the time, we had about fifty females on station assigned to various functions throughout the squadron. Once again, the visitor seemed very bored with it all. Following the base tour, I took the VIP to lunch at the dining hall. I had set up a table in the commander's dining room for her and twelve other squadron members representing most of the functions on base. I had our female non-commissioned officer-in-charge (NCOIC) of the supply, services, and administration department seated next to the VIP. She had very little discussion with the airmen and again seemed bored by it all.

Prior to her departure she asked if she could get a tour of the city of Galena. Since we had about two hours before departure that would be no problem. I took her the long way to the New Town of Galena along the dike road through Old Town Galena. She seemed interested in what the locals did for a living. Just as we entered Old Town, we observed one of the natives taking a leak from the front porch of his home. She asked if that was normal, and I said yes it was and that it was a good thing that he wasn't doing number two. She was so surprised as she had so many positive impressions of the local people taking care of their environment.

The next observation completely blew her mind as she saw mounds and mounds of trash sitting in the middle of the frozen Yukon River. There were old cars, snow machines, refrigerators, animal carcasses, and other garbage piled high in the middle of the river. I explained that this is how the Athabaskan people did spring cleaning in Galena. They piled all their unwanted trash on the river and then let the river take the garbage out to the Arctic Ocean during the river breakup cycle which would probably occur later in the month. So much for honoring the environment.

Break-Up Begins. One event that happens each year is the breakup of the Yukon River. Depending on the thickness of the ice and the prevailing weather, the breakup can be a very minor event or a catastrophic event like the beak up of 1989. The Yukon River is such a large river; no words can accurately describe the scale of things along its more than 2,000-mile length. The Yukon, as it passes by Galena, is nearly one mile wide. There are sloughs running through large wooded islands that are larger than most rivers. The current in the river is very powerful, and you really need to know how to navigate the river. The headwater of the Yukon is in the Canadian subarctic and flows across the entire breadth of Alaska to where it meets the gigantic delta in the Bering Sea.

One of the biggest challenges I faced during my command at Galena was dealing with the river breakup of 1989. While there were a few very outdated checklists in place to prepare for the breakup, I decided early on that the squadron should review all the items in the checklist, verify the validity of the recommended actions, and practice each item on the checklist so we knew exactly the steps to take in case of flooding from the breakup. I directed the staff to begin the review of the breakup checklist on May 1, and we practiced each functional area of support three times per week. We found many deficiencies in the checklist and fixed each deficiency as we proceeded. The actions in the checklist were predicated on the level of the river during the break-up process. We had a very well-defined river flood gauge on the east end of the runway and each action we took was dependent on the level of the river. The static reading on the gauge while the river was frozen was at the 120-foot mark. The flood gauge gave us an immediate reading to determine not only the depth of the river rise but also the rate of rise.

Some of the practice actions we took involved moving critical resources to the top of the dike that surrounded the base, determining what equipment would be at risk if the flood waters rose to a certain level, and also determining the critical number of core personnel that would remain at Galena if the base had to be evacuated before the flooding closed the runway. One of the most irritating, to the airmen at least, practice events that took place was to require all the personnel that would be evacuated in case of the flood to show up at the flight terminal with the personnel belongings they would require during at least a two-week evacuation. My first sergeant developed a critical equipment and personal article list for the airmen to use for an evacuation. During the practice evacuation we required each person designated to evacuate to come to the air terminal with their personal items they would take with them. The first sergeant inspected each bag brought to the terminal and determined that a few of the folks did not take the exercise seriously because they had filled their travel bags with dirty laundry and bedding. Each person that did not have the correct gear got a good butt chewing from the first sergeant. In addition, our operations staff filled out a practice passenger manifest for the 270 folks that may need to be evacuated if the river flooded.

Just as important, we fine-tuned the list of key personnel that would need to remain behind to "save" the base if the flooding became severe. After much discussion and compromising we determined that there would be fifty-one key personnel that would remain behind at Galena if there was a flood evacuation.

May 7, 1954 Hours. "Notified at my quarters that the ice on the river officially broke this evening." After penning that journal entry, I drove to the river's edge and noted the river level was at 125 feet. Water was flowing on the edges of the river but there was little ice movement. I immediately recalled the staff and had our first real coordination meeting in preparation for possible evacuation. I positioned two security policemen near the river flood gauge with instructions to report any rise in the river level. We had the first increase in the river level at 0200 hours on the morning of May 8 to 135 feet. I directed each functional area to conduct a real-time personnel head count to determine the number of personnel actually on base. The supply and services sections could not get an accurate count. We needed to have an accurate count to ensure we had enough airlift support in case we had to evacuate the base.

May 8, 0800 Hours. When I reconvened the staff meeting, the gauge reading on the river was at 137 feet and eleven inches. I notified the Elmendorf command post of the river's rise rate and requested that they put the evacuation aircraft on standby alert. I next directed my staff to finally complete an accurate personnel head count of each functional area. Since we had a C-12 aircraft on base from a flight the previous day I requested that the crew fly me over the river to get a good visual reading. Massive chunks of ice were slowly moving to the west with some large build up on the north bank of the river. There seemed to be an ice blockage about twelve miles downriver at a location known as Bishop's Rock.

May 9, 0830 Hours. At the morning staff meeting I was notified that river had risen from 137 feet to 140 feet in the past ninety minutes. I received a call during the staff meeting that the 21st TFW commander was going to fly an F-15 to Galena to check on the river. I advised him that was not a good idea as the 140-foot river level was the cut off level to begin evacuation procedures. The wing commander took my advice and cancelled the F-15 flight. However, the Alaskan Air Command commander, a three-star Air Force general, and the wing commander notified me that they would be flying to Galena in a C-12. The 140-foot level also activated the process of driving key vehicles onto the dike that surrounded the base. All of the key vehicles and support equipment were in position on the dike by 1130 hours. I then contacted the Alaska State river experts and they advised me that we had a 90 percent chance of reaching the 154-foot mark in less than eight hours based on the rate of rise in the river to this point. The 150-foot level would breach the dike around the base and most likely flood all the Air Force buildings inside the dike. The AAC commander and the wing commander landed at 1500 hours and asked for my recommendation on evacuation. Based on the breakup checklist that we used and the statements by the Alaska river experts, I told the general that it was time to evacuate personnel from Galena to Elmendorf. The river level when I made the decision to evacuate was 141 feet, eight inches and slowly rising. The first of the eight C-130 aircraft arrived at 1945 hours and by 2400 hours we had all 270 evacuees airborne en route to Elmendorf. Our pre-evacuation planning really paid off as we had no problems processing the evacuating airmen. At 1645 hours I had ordered the launch of the two alert F-15s to deploy to Elmendorf and maintain alert status there until they could safely return to Galena. The weather was very cold at 14°F and very windy. The remaining fifty-one key personnel met in the commander's conference room, and I gave them instructions to just follow the evacuation checklists and report any problems that they noted. I kept the two security policemen near the river flood gauge to give us continual real time flood level updates.

May 10, 0800 Hours. Reconvening the staff in my conference room, I directed our key personnel to walk through the next potential steps in the evacuation process: to evacuate the key personnel and abandon the base. We surveyed the location that the rescue helicopters could hover and pick up the remaining fifty-one personnel from Galena and fly them to high ground. I was envisioning being the last person off the base but was not going to go down with the ship.

Since there was the very real possibility that the Galena power plant would go down due to flooding my services officer recommended that we attempt to clean out all the remaining frozen and perishable food products in the dining hall, the club, and bowling alley snack bar. Our head cook came up with some very incredible meals for the remaining "survivors." We had the most amazing steak, scallop, and fried chicken meals. The river remained at 141 feet, ten inches for four hours – only nine feet below the flood level that would completely overwhelm Old Town Galena. Once the perishable food was eaten, we reverted to the standard Meal Ready to Eat packages (MREs). Fortunately, we had nearly 135 cases of MREs on hand that would get us through a least a ten-day period.

May 11, 0800 Hours. The water level was steady at 141 feet, and my initial thought was to start the coordination process to get the evacuees back from Elmendorf. The biggest shocker came when I got a call that the Alaskan Air Command inspector general (IG) was going to continue with their plans to conduct the annual unit effectiveness inspection (UEI) of my base starting on May 16. You could almost guess my reaction to this call. I immediately called the 21st TFW commander and told him "You gotta be shitting me! I have 270 of my airmen deployed due to the flooding of the Yukon River, my mission is being done at a different base, my base is partially flooded, and the IG wants to come and evaluate my squadron in one week? I don't even know when my deployed troops will return to Galena." His response to me was, "Don't worry, you will do fine in the inspection!"

I went flying that day with the Alaska Division of Emergency Services to get an eyeball on the river. The river was an awesome mass of ice slowly moving to the west. There appeared to be a very large ice jam about seven miles downriver. When I landed the river had risen again to 143 feet, seven inches. Just as I arrived back at the office, I had a call from the governor of Alaska's staff asking if it would be OK to utilize the A-10 aircraft from Eielson Air Force Base to bomb the ice jam down river and open up the flow of the moving water. I told the governor's office to stand by on that request and I called the Alaska river experts with the governor's concept. They told me that was the dumbest idea they had ever heard and artificially clearing the buildup of ice could have additional catastrophic results; they also said that the ice flow would eventually clear itself on the way to the Arctic Ocean. I called both to the governor's office and the wing commander's office telling him the governor's idea was not a sound one and that I highly recommended not bombing the ice jams. Fortunately, clearer heads prevailed at the governor's office and they dropped the idea of bombing the river with the A-10s.

May 12, 0700 Hours. Day six of the evacuation started off with the staff meeting, and I was briefed that the river had risen again during the night to 148 feet, eleven inches. I then decided that we would delay the return of the airmen from Elmendorf another twenty-four hours with the hope the river would eventually recede to a safe level. I flew again over the river and saw some additional movement of the ice, especially along the north bank, closest to the Galena runway. The ice jam near Bishop's Rock had started to break up. When I landed in the late afternoon the river level had dropped to 145 feet, eleven inches. The ice buildup on the sides of the river was absolutely awesome as ice chucks the size of cars were being moved around by the flowing water as if they were just toys.

May 13, 0600 Hours. Most of the night was spent waiting for updates every two hours, but the river level maintained at about 148 feet throughout the night plus or minus a few inches on each river check. Early in the morning I notified the 21st TFW vice wing commander at Elmendorf that we could start the process of bringing the airmen back to Galena on Monday, May 15. He requested that we wait another twenty-four hours so the wing support staff would not have to work on Sunday preparing the return of our airmen. That decision was OK with me as we still had a few things to do to get ready to receive our airmen back at Galena and that twenty-four-hour delay would give the river some additional time to recede more. I talked to my operations officer back at Elmendorf and he said the squadron morale was good, but they are all ready to come home. They were calling themselves the Yukon Refugees. Those of us left at Galena were calling ourselves the Yukon River Rats. Just as the day was coming to an end, I got a call that the river had unexpectedly risen to 150 feet, three inches because of a huge ice jam at Nulato just down river from Galena. The river was now coming over the dike and had caused significant flood damage to Old Town Galena and the Galena dump, and was approaching the electrical power plant on the west end of the base.

May 14, 0500 Hours. Based on the rapid rise in the river I called for the daily staff meeting at 0500 hours to deal with the flooding that was now occurring because the dike had been breached. The electrical power grid to the west end of the runway was rapidly starting to take on water, and I ordered the civil engineers to shut off all power to that section of the base including power to the runway and a portion of the alert hangars. Our base dump, the treated water outlets into the river, and the southwest side of the dike were now underwater. I was concerned that the flooding would degrade the material holding the dike together. We needed to move some of our equipment and vehicles off the southwest side of the dike to preclude them from falling into the flood waters if the dike gave way. The rumors started to come fast and furious from Elmendorf that Galena had been lost, the runway had washed away, and all the remaining airmen at Galena had been evacuated. I contacted my operations officer at Elmendorf and told him to gather all the Galena airmen at Elmendorf and give them the true picture of what was happening at Galena and that I would get them back to Galena just as soon as we could. By the end of the day the water level had dropped to 148 feet and was slowly receding.

May 15, 0600 Hours. The morning gauge reading showed that the river had stabilized at 148 feet and was slowly going down. I made the call at 0930 hours to launch a C-130 from Elmendorf with some of the runway control tower operators, approach control radar operators, weathermen, and aircraft maintenance troops to recover the inbound aircraft returning the Galena "refugees." I also coordinated the return of the remaining evacuees for May 16. The first C-130 aircraft landed uneventfully at 1315 hours. While I was at the base operations building welcoming our returning airmen, I got a call from the State of Alaska emergency services director requesting that I delay the return of the deployed personnel indefinitely as they were concerned about our base's treated water flowing into the river because of the flooding of our normal over flow tubes. I immediately called back to the wing commander and asked him to coordinate with the state to cancel that request. My local civil engineering officer that stayed behind during the evacuation told me that there was no problem with our wastewater flow after the flood. Fortunately, this issue was not a problem and the state folks called back with approval to return the airmen. I told them that was nice but that they didn't have a vote in the decision; my airmen would all be returning to Galena the next day.

May 16, 0700 Hours. The river continued to drop, and we are now at a steady 146 feet. I called the 21st TFW commander and gave him the "green light" to launch the remaining C-130s to return the rest of our airmen to Galena. The first of the seven C-130 flights arrived at Galena at 1330 hours, and the last flight arrived at 2230. Finally, all 270 deployed airmen had returned back to Galena. I greeted each returning aircraft to welcome the airmen back home. The morale seemed high but the base was in complete chaos as we worked to put our functional areas back together.

Congressional Inquiry. While I was very busy coordinating the arrival of our returning airmen, I got a call from Washington, D.C. about a congressional inquiry that had been filed against me regarding a civilian employee I had fired recently for a very vile racial slur he made against one of my sergeants. The letter that was filed by the civilian employee and stated that I was a young, inexperienced commander who was against all Vietnam War veterans. I politely told the caller that I myself was a Vietnam War veteran, that I had flown 535 combat sorties during the war, and that I would not allow anyone in my squadron to call anybody a "nigger." In my mind this case was closed, and I never heard anything else from the congressional office about the firing.

Unit Effectiveness Inspection (UEI). On May 16, I was notified that the Alaskan Air Command inspector general and his thirty-member inspection team would arrive on May 22 to begin their inspection of the squadron. I tried protesting to the wing commander with no avail, the inspection was coming. We had loads of things to put back together to prepare for the inspection. The Air Force unit evaluation program was a well-defined inspection program to do the following:

"Evaluates the leadership effectiveness, management performance, aspects of unit culture and command climate, and the ability to minimize waste and prevent fraud and abuse.

"Provides the Secretary of the Air Force, the Chief of Staff of the Air Force and all commanders at all levels an independent assessment of unit compliance with established directives.

"Enables and strengthens commanders' mission effectiveness and efficiency through independent assessment and reporting of readiness, economy, efficiency, state of discipline, and the ability to execute assigned missions.

"Motivates and promotes military discipline, improved unit performance, and management excellence throughout the chain of command and within units and staffs.

"Identifies, reports, and analyzes issues interfering with readiness, economy, efficiency, discipline, effectiveness, compliance, performance, surety and management excellence.

"Supports and informs commanders' risk management at all levels. IGs must ensure the Air Force Inspection System supports prudent decisions by commanders that have documented accepted risk.

"Enables Functional Area Manager assessment of functional effectiveness, field compliance, and of the adequacy of organization, policy, guidance, training and resources.

"Provides a mechanism for Air Force senior leaders to direct a targeted, detailed, and thorough inspection of specific programs, organizations, or issues.

"Reinforces to commanders and airmen at every level the quality of mission readiness and inspection readiness.

"Eliminates on-site inspections which are not mission-relevant, do not outweigh the cost, or detract from mission performance and readiness.

"Significantly reduces (with the goal of eliminating) the wasteful practice of "inspection preparation."

As you can see, this UEI was going to be a big deal. It essentially was my report card on how well I was leading the squadron at Galena. I had been able to delay the unit inspection from April 16 until May 22 because of the massive amount of damage we had from the freeze in January and the flood evacuation of 270 members of the squadron for eight days in early to mid-May. We still had lots of preparation to do to get ready for this most important inspection.

Thieves on Every Corner. Just as the final days of preparation were in overdrive, I had to deal with three separate issues of theft on the base. The first one dealt with one of my civilian employees who was caught earlier in the month stealing government gasoline from our fuel shop. He had filled five 10-gallon containers and was preparing to leave the base with the fuel. I was notified by the security police that they had caught him in the act of stealing the gas and then stopped him as he was trying to depart the base. As soon as I was briefed on the incident, I immediately barred him from base and began the tedious task of firing him from civilian employment. Right on schedule, after the morning staff meeting, I got a call from the civilian employee's wife in the City of Galena begging me not to fire her husband. Based on the fact that this employee had numerous other work-related complaints, I told her that I was going to go through with the firing because I just could not accept thieves on our base. She was none too happy and advised me she would be contacting her U.S. representative and filing a formal a formal complaint against me. Since I had just dealt with another congressional compliant, I had the contact information available for her to use to contact her congressman. I never heard another word about this incident from her. After nearly four months the civilian personnel system approved my decision to fire the employee.

The next case I dealt with this very week was the theft of a fairly valuable watch from the junior dorm. I was advised that the person that "lost" the watch found another airman wearing the watch the very next day. I asked the security police to look into this incident and found out later that day that the watch was in fact stolen; the security police confirmed that the rightful owner of the watch had his initials engraved on the back it. In spite of the thief claiming it was his watch, the evidence was overwhelming, and I brought Article 15 charges against the thief. His punishment was a $150 fine for two months and a one-month base restriction. Oh, the joys of command.

While the staff was removing all the critical vehicles and equipment from the dike following the flooding, they found that two brand new Evinrude 65-horsepower outboard motors were stolen off two of our Morale, Recreation and Welfare (MWR) boats. The cost of these two motors was approximately $3,200, so this was not a minor theft. The boats were positioned on the dike to cover a last-ditch effort to save the personnel who remained on the base during the flood if the rescue helicopters could not evacuate them. Our security police contacted the Galena Police Department to advise them of the theft. After a two month search the local police believed the outboard motors were transported to one of the outlying villages on the Yukon. The motors were never found during my time at Galena.

UEI Preparation. Each functional area continued with the preparation for the upcoming UEI. Two of my key personnel, the operations officer and the chief of maintenance, were scheduled to leave Galena before the UEI began. Both were past their normal departure dates as they had been deployed to Elmendorf for the flood evacuation. That would leave two huge holes in the leadership roles for the UEI. I made it a point to visit every functional area more as a cheerleader than a commander. I could certainly feel the pressure as each function on base not only tried to recover from the January freeze and the May flood evacuation, but also fine-tuned the inspection items that would be reviewed in less than a week.

UEI Begins. The IG team arrived on schedule on Monday, May 22 with the Alaskan Air Command inspector general, a full colonel, and thirty other functional inspectors. My team's UEI in-briefing was flawless and all the support items were in place to assist the inspectors. It is always good to start off on the right foot, and we gave the inspectors a positive first impression of the unit they were about to inspect. During the UEI in-briefing, the wing commander at Elmendorf ordered a practice scramble of our two alert F-15s. While it was just a practice scramble, the jets got off right on time and our maintenance airmen's support was exactly by the book. Unfortunately, one of the jets returned with a major mechanical issue, and we had to put the jets on mandatory scramble orders (MSO) while we awaited the replacement part from Elmendorf. The aircraft was repaired and back on alert status by 2230 hours that evening.

During the next day when I was escorting the head of the IG team around the base, we got an Amber Warning to immediately scramble the F-15s. They flew out to the northwest of Galena and intercepted two Soviet Tu-95 Bear G bombers. The launch, scramble, and maintenance support for the launch was flawless and noticed by the inspectors. However, despite the flawless scramble both fighter aircraft had mechanical issues and had to return to Elmendorf for repairs. The jets were fixed and were back on alert status at Galena by 2100 hours.

After the normal duty hours, the IG requested that I meet him in his quarters to discuss the progress of the inspection. The entire inspection team leadership was there and they gave me some very positive feedback: the inspection was going well and that they had, so far, not found any major issues. The IG then threw me a real curve at me and asked if I would take an assignment back to Elmendorf when my tour was complete in October to become the Alaskan Air Command's director of inspections. I really didn't have an answer since I didn't have a follow-on assignment yet, but I did not turn him down. I certainly did not want to jeopardize the results of this current inspection by directly turning down the IG's request for me to join the IG team.

Finally, it was judgement day from the UEI. We gathered the entire squadron of those personnel that were not actively working to assemble in the base theater for the UEI out brief. I had mixed feelings about the outcome as I was aware of some minor problems in the ground support function and the supply functions. It was nothing major but maybe it was enough to keep our overall score down. The inspector general came to the podium and announced to the entire squadron that the overall rating of the 5072nd Combat Support Squadron at Galena was Excellent. The theater erupted in shouts of joy and relief. The airmen came through when it really counted. I was ecstatic and even more so when I saw the written results of the inspection. The executive summary of the formal inspection report from the IG follows:

"Lt. Colonel Thomas M. Petitmermet was providing strong, innovative leadership to the men and women of the 5072nd CSS. He was the proactive catalyst behind significant improvements, superior people programs, and the highest unit morale ever seen. His hands-on approach to leadership provided positive results across the entire spectrum of the unit. Effective communications, and recognition when necessary were the cornerstones of his success. Colonel Petitmermet was deeply involved and concerned with personnel issues and unit problems. His Commander Gram program provided invaluable feedback concerning unit operations and weaknesses. He truly was a commander with his finger on the pulse of his unit's readiness to accomplish the mission. Galena never looked or operated better."

Wow, this was my report card during my time as the commander of the 5072nd Combat Support Squadron at Galena. We could never have pulled off this rating without the incredible teamwork and dedication of each and every member of the squadron. Despite the many challenges the squadron faced, we came through with flying colors.

Time to Unwind. I declared Friday, May 26 a unit down day. Our MWR folks set up a squadron-wide sports day with lot of activities going on. The noon meal was a very well planned and executed outdoor barbecue with hot dogs, hamburgers, sides, and gallons of beer. I managed to play three softball games – I was seven for nine at bat – but I really stretched every muscle in my old body. After the events were complete, I went around to every person competing and every work location that was carrying on the mission to thank them personally for their stellar performance during the preparation and conduct of the UEI.

Assignment Anxiety. It was now less than five months before I would move on to my next assignment in the Air Force. I was starting to get concerned that nothing concrete had developed. I had been asked by the IG to take the director of inspections position on the Alaskan Air Command IG team. I was not certain that I wanted to have another assignment in Alaska as the Galena assignment was my fourth job in The Last Frontier. I had an earlier request from a former boss of mine, a one-star general, who wanted me to come to the North American Aerospace Defense (NORAD) staff. That certainly was a possibility. Also, my wing commander told me earlier in the month that I should take an assignment to the Pentagon in Washington, D.C. to enhance my possibility of getting promoted to general. The Pentagon was definitely not an assignment that I would try for. One of my career goals had been to never be assigned to the Pentagon. At this point in my career, twenty years, I didn't want to interrupt that goal. So, there was more waiting to find out where my family and I would be moving to in October.

May Intercept Summary
May 23: 2 Tu-95 Bear Gs

Chapter Eleven
PERSONNEL PROBLEMS ON STEROIDS
June 1989

Drunk on Duty. I'm not sure if it was the warmer weather or post-UEI let down, but June started off with some very intense personnel problems. On June 3 I was notified that one of the young security policemen showed up to his duty station still drunk from the night before. I was very concerned that the chief of security police did not take any action, like get a blood alcohol test (BAT) on the drunk airman. His excuse was the Air Force BAT tester was unreliable. So, being the bad old commander, I directed that the airman be sent downtown to the Galena police station to get a valid BAT test. Five hours after reporting drunk to work his BAT was .178, more than two times the legal limit in Alaska. Thankfully, he did not drive a vehicle to work or drive a vehicle after reporting to work. I sent for the chief of security police, and we had a very long, detailed meeting on this issue. I advised the chief that I would process Article 15 action against the drunk airman. The chief was reluctant to agree to that course of action because he felt it would affect the morale of the security police force. My answer was, "Tough shit. He is getting an Article 15." Not less than two weeks later a second security policeman showed up to his duty station very drunk. I was obviously very concerned by the leadership of the security police. I asked the top three leadership positions in the squadron to come by my office to help develop a plan where we wouldn't have any more drunk airmen showing up to work. All three were very receptive of a plan to more closely monitor the young airmen in the security police section by engaging the middle level non-commissioned offers (NCOs) to be more aware of the drinking problem at Galena. For the most part they did their duty, but another very serious drinking issue arose from the security police section.

Under a Watchful Eye. Late one afternoon a young female airman from the base administrative section came by my office after 1700 hours. I was the only one left in the office, and she told me she waited until everyone had left the building so she could talk to me alone. She told me that a new technical sergeant from the services section who had just arrived on the base told her that he was an undercover Office of Special Investigation (OSI) informant and that he was sent to Galena to spy on the commander, me, because of many unlawful personnel actions and financial transactions that I had done. She was so concerned by the story that she thought I should be aware of the situation. I immediately called the OSI office at Elmendorf and relayed the story about the OSI "informant" sent to Galena to find "dirt" on me. The lieutenant colonel in charge of the OSI at Elmendorf told me he was unaware of any undercover activity at Galena but that he would get back with me in the morning. I was left wondering if there really was an informant at Galena but the OSI official at Elmendorf could not tell me if I was under investigation.

Not waiting for this issue to get out of control, I called my chief of security police to my office at 0700 hours to get a handle on how we should proceed with this issue. During my meeting with the security police chief I got a call back from the OSI chief at Elmendorf stating that there was no official or unofficial OSI undercover investigation going on at Galena.

I next asked to see the "informant's" personnel records so I could get a feel if this really could be an undercover operation. This sergeant in question had a very mediocre record of performance over the eighteen years he had been in the service. I called back to his previous unit and talked to the services squadron commander to which he was assigned. The commander relayed that this individual was a real troublemaker; the commander was surprised this sergeant was allowed to transfer to Galena because he had made similar claims of being an OSI undercover informant in 1981 and 1986. This BS had to stop immediately.

I called the sergeant to my office later that afternoon and directly confronted him about his claim of being an OSI undercover informant. He said he was just using that line to pick up "dates" from the women in the squadron and that he was just here to do his job in the services branch. I asked him about the two previous complaints against him, but he really never gave me a satisfactory answer. I then directed my chief of security police to further investigate what the hell was going on with this sergeant.

Over the next two days the sergeant wrote an eighty-one-page rambling essay about what his mission in the Air Force was and that he really was an undercover OSI agent sent to spy on me. I had enough of this crap and directed that he immediately be sent back to Elmendorf for a mental health evaluation. After a twenty-four hour observation, the mental health giants at Elmendorf advised me that there was no problem with the sergeant, except that he had just had drank too much coffee the past three days and would be just fine to come back to Galena and continue his Air Force duties. Needless to say, I was infuriated with this decision. This was the second time in three months (remember the young plumber) that the mental health experts at Elmendorf sent a very mentally disturbed person back to the squadron.

Obviously, we needed to come up with a plan to deal with this returned sergeant. I met with my first sergeant and we came up with a plan to very closely monitor this sergeant once he returned to Galena. I started off with a very direct face-to-face meeting with the sergeant telling him that we would be watching him very closely. I gave him a direct order to drop his "story" about being an OSI undercover agent. I issued him a Letter of Reprimand for his lying about being an OSI agent and advised him that one more, even minor, indiscretion would result in me processing him for separation from the Air Force. I next had his direct supervisor and the officer in charge of his section in my office telling them what our plan was to deal with this troubled sergeant. During my remaining time at Galena, we had no further problems with this individual.

Marriage Counselor. One of the more unique personnel challenges I faced at Galena involved a twenty-one-year-old sergeant who wanted to marry a local citizen from the village of Galena. The question came to me because the young sergeant had requested permission to let his new "wife" stay with him on Galena for the maximum fifteen-day period and then let him move off base to live with her in Galena. My housing chief brought the request letter to my attention. I was obviously concerned about this situation and called in my chaplain, first sergeant, and chief of security police to get some advice. My security police chief strongly advised me to disapprove this request. It seems that the local civilian that my sergeant wanted to marry was a fifty-six-year-old woman who had three grown children and seven grandchildren who all still lived in the village of Galena. This woman had previously tried to marry three other Galena airmen in the past with the explicit intent of getting her entire family on the government dole. By becoming a dependent of the sergeant, her entire family could become eligible for medical, commissary, base exchange privileges, and many other government benefits. My chaplain also said that this would be a very bad move for the sergeant and could really foul up his future.

I next called in my personnel chief and requested information about when this young sergeant in question was due to finish his one-year assignment and depart Galena. The sergeant already had a follow-on assignment to Nellis Air Force Base in Las Vegas, Nevada. Fortunately, the sergeant was due to leave Galena the following month. I enquired about moving his departure date up a few weeks since he already had orders Nellis. The personnel chief said it would be no problem to move the departure date up by two weeks.

I discussed my options with the first sergeant and the chaplain, and they both agreed that the best course of action would be to deny the visitation and off-base move request and then process the sergeant for early departure from Galena. The off-base question was no problem, as no active duty military were allowed to live off-base during their one-year remote assignment. I felt the right thing to do was to bring the sergeant into my office and advise him of my decision. Just two hours later the sergeant was in front of my desk hearing my decision.

He was not too pleased with what I told him, but I gave him as much fatherly advice as I thought he could handle. He was completely unaware that if he married this woman her entire family could become his dependents. How could he manage a family that large? What would that do to his very promising career in the Air Force? I told him that I was not going to approve his on base visitation request, nor his request to move off base after his marriage. I also informed him that I was moving up his departure date from Galena by two weeks to help in his decision process to not marry this local woman. While he told me that he was deeply in love with the local women, I sensed that he showed a deep sigh of relief that I had saved him from a potentially life-changing scenario.

PMS Problems. I title this section PMS problems (post-menstrual syndrome) as I had to deal with two very nasty older female senior sergeant issues. The first involved the senior master sergeant in charge of our supply section. Of all the reports from the recent UEI, the supply section had the most negative write ups and the most problems to be addressed. One of the action items in UEI report addressed the near marginal control of some of our supply items. The narrative part of the report also addressed the very negative leadership of the supply chief. This is the very same senior master sergeant who had been giving the entire squadron problems with all her unfounded racial accusations against anyone who disagreed with her. The UEI write-up implied the same and mentioned the terrible morale in the supply section. I really had no choice but to fire the supply chief and place her senior technical sergeant in charge of supply. I felt that I had no choice but to present my findings and those of the IG team during my talk informing her that she was no longer in charge of the supply unit and she would be, at least until she left Galena, working directly for the first sergeant doing whatever tasks he needed accomplished. Her response was immediate: she had already filed a social actions complaint against me and wrote to her congressman, and then gave me a list of things that I must do before she left Galena for her next assignment. Among her demands was a nine (highest rating) on her efficiency report, the award of a meritorious service medal, and an apology for my racism in writing. I nearly came out of my chair and choked her. Fortunately, I had the UEI written report and the notes I had compiled over seven months documenting her poor performance up to that date. I knew I was covered and told her to get out of my office and report immediately to the first sergeant. During the next month I heard continual complaints about her, but she finally left Galena, and I never heard another thing about the formal complaints that she said she was going to file against me.

The next female problem came from the master sergeant in charge of billeting at Galena. Billeting had a satisfactory rating from the IG, nothing terrible but nothing excellent or outstanding. One item from the IG report was a complaint filed by this master sergeant accusing her boss, a captain, of some very serious wrongdoing. I immediately requested that 21st TFW start an investigation to determine what these accusations against services captain were all about.

Normal procedures at Galena was for the commander to approve a DEROS (Date Estimated Return from Overseas) or the date an individual would depart Galena for their next assignment. This DEROS usually occurred a few days prior to the one-year anniversary of their arrival at Galena. Individual departure dates depended on the departing individual having their section in order and a thorough transition briefing to their replacement. In most cases this date would be a few days, at the most, before the one-year anniversary of their arrival at Galena.

I got a call from the 21st TFW vice wing commander early on the morning of June 2 saying he had gotten a call from the husband of this master sergeant cussing him out because I would not let his wife leave on her DEROS flight on June 3. Her one-year anniversary of arriving at Galena was June 27, a full twenty-four days after her requested DEROS date. Since the 21st TFW was still investigating her complaint against the services captain I needed her to

Chapter Eleven 123

remain at Galena until the investigation was complete or at least progressing as she was the one and only complainant and witness to the alleged problem she presented to the IG.

Being a person of action, I called her into my office and told her that she could not depart Galena until the 21st TFW had interviewed her and corroborated her accusations. You would have thought that I had just killed her first born. She was now going to file an IG complaint against me for holding her at Galena against her will and doing it only because I was upset. I advised to her to go ahead and file the complaint, but she was not leaving Galena until the wing was satisfied that they had the all the necessary details and had investigated her complaint. Almost as I anticipated, the wing completed the investigation by June 10 and found no facts to support her complaint. Her complaint was dismissed, and I sent her on her way out of Galena on June 12.

No Help from Headquarters. The buffoonery from our Air Force Military Personnel Center (MPC) at Randolph Air Force Base, Texas really came to a head in June as there were some very troubling assignment actions from them. First, we were sent an eighteen-year-old Airman Basic right out of basic training to a one-deep position in our chaplain's office. She was motivated but had no clue on what she was supposed to be doing. Our one chaplain had his hands full enough just trying to keep up with all the personal issues chaplains usually deal with. We had to request additional manning from the 21st TFW to help manage the chaplain's office.

We had three last minute assignments cancelled for people that were due to report during June into some very key positions. It was difficult trying to keep those key functions working when we did not have the minimum number of people assigned. The folks in those functional areas really had to pick up the slack. I was happy that the UEI was behind us but was still concerned as we were facing an upcoming operational readiness inspection (ORI) with some key personnel slots not filled. This ORI was scheduled only two months after our very intense UEI.

We also had the MPC personnel folks "change" three out-bound assignments of some of our people after they had already had their personal goods moved to the new base; in one instance, the individual had already moved his family to the new location on the authority of orders he had received months earlier.

I was none too happy with all of this personnel baloney and fired off a very direct, in your face, message to the Military Personnel Center at Randolph Air Force Base expressing my utter disdain for their continual screw ups of my squadron's personnel assignments. I had provided a copy for my wing commander and followed up with a phone call to him. It wasn't twenty-four hours later that I got a call from a very irate personnel colonel at Randolph telling me to quit complaining about the personnel assignment process. That call didn't go very well for him, as I let him have it with both barrels and described each assignment screw up that his people had done. I also reiterated that at a remote base it is very difficult to just make do with what we have, as the original manning documents for our squadron laid out the very minimum manning levels we needed to accomplish our mission. I was so forceful with the colonel that I was expecting that my upcoming assignment would be a terrible job to some far-off location in the world. I don't believe the personnel issues were ever fixed to my satisfaction as it seemed like every month we were dealing with the personnel center incompetence.

Real Bear Hunting. As flood waters receded, we had a real wildlife challenge one day. The Galena dump was located about 300 yards from the western edge of the base perimeter. It was not unusual to see black bears rummaging around in the dump for food following their hibernation period. I got a call one afternoon that a fairly large black bear had somehow entered the base and was near the combat alert cell (CAC) causing the detection monitors to sound an alarm when the bear walked near the alert hangar. We had electronic "eyes" on each side the alert hangar doors, and this bear was causing the alarm to activate each time it blocked the view of the opposing electronic eye. My security police experts wanted to shoot the bear because it was causing a real problem around our alert aircraft. I told them to stand by while I called the local fish and wildlife officer in Galena City to describe the situation. To my surprise he told me that if the bear was a threat to the aircraft or the personnel at the CAC to go ahead and shoot it as long as we could do it safely without endangering any facilities or people. The security force was ecstatic, and they quickly dispatched the bear with multiple M-16 shots. One of my maintenance airmen, who came from West Virginia, said he knew how to dress out a bear and we could then cook and eat the meat. Bad idea. That was the worst tasting, nastiest meat I have ever eaten.

Chink in the Armor. Galena is one part of a very elaborate and expensive system known as the North Warning System designed to protect the U.S. and Canada from any potential Soviet aggression over the North Pole. Part of this system is an array of FPS-117 radars that I described earlier in the book. These radars are placed in strategic locations along the frontier of the U.S. and Canada, and Galena was host to one of these FPS-117 radars.

During a morning staff meeting in early June, the civil engineer officer brought up the concern of the FPS-117 radar contractor at Galena that there seemed to be an intermittent glitch in the power supply that was being used to support the radar. Some of the electric power to the Galena radar system came from the City of Galena with backup power provided by the four very large power generators on the Galena base. There was an obvious problem somewhere along the power grid to the radar. My civil engineer and I travelled to the City of Galena to discuss this power issue with the city folks. The city engineer had no knowledge of what could be interrupting the power supply to the Galena radar. Following our meeting in the city, as we were driving back to the base, we noticed large areas of the landscape to the east of the base that had been washed away by the recent Yukon River flooding. We stopped our vehicle and noticed that there were several large moose feeding in the area where the flooding had taken place.

My civil engineer knew that the power cables from the City of Galena passed near the area where the moose were feeding. We got out of the vehicle, walked to that area, and found that because of the flooding almost all of the soil that was covering the power cables had been exposed, and the moose were chewing on the exposed power cable. Luckily, the moose had not completely chewed through the cable or we would have had some "fried" moose on our hands. We found the source of the power problem; the cable was repaired and reburied in the ground. Such an innocuous situation could have brought down a key link in our North Warning System.

June Intercept Summary
    Zero Intercepts

Chapter Twelve
MORE PERSONNEL PROBLEMS
July 1989

Rape Charges. July started off with a "bang." As I was just finishing up my annual two weeks of leave back at Elmendorf, I got a call on July 2 at 0530 hours from my vice commander that there had been an accusation of a rape the previous evening. It seems that one of my services sergeants had accused one of my security police sergeants of raping her after a wild party at the barracks. I gave my vice commander instructions to get the security police involved and have them start gathering info on the alleged crime. I must state right now that I had the very best vice commander in the Air Force supporting our mission. Every time I was TDY or away from Galena, my vice commander took care of every single problem with perfect results each time. It was a relief to have such a professional officer on our team. My vice commander had directed that the two sergeants involved in this situation stay away from each other while the wheels of justice progressed with the rape investigation.

The work on this rape case took several months to come to a conclusion, and I will describe the outcome of the case in the September summary. Needless to say, this rape case was just another one of those 10 percent events that took up 90 percent of my time.

No Notice DV Visit. During early afternoon of July 10 I got a call from our base operations office that we had a "code" on the normal rotation aircraft that flew to Galena and King Salmon twice a week. The rotation aircraft was a contracted aircraft flown by Reeves Air in a Lockheed L-188 four engine aircraft. A code is a term which designates that a person in the Distinguished Visitor (DV) category was on board the aircraft. It could be anyone from a 21st Tactical Fighter wing colonel to a four-star general or equivalent. I immediately went to base operations to find out who the code was since we had no prior notification on the flight plan. My operations coordinator told me the DV was still on the airplane and that he wanted to see the base commander. Oh, what have we done now? I boarded the aircraft and saw an Army four-star general sitting toward the back of the aircraft. I approached him and apologized that we didn't properly meet the aircraft but that we did not have any notice that he was on board. He said, "No problem, Colonel. I told the crew not to announce my presence on this flight." He then stated that he needed me to get this aircraft refueled as soon as possible as he was on his way to King Salmon, the other F-15 alert base in Alaska, and he wanted to start his salmon fishing trip as soon as possible. I told him I would take care of that and get him on his way as soon as possible. I noticed his name tag was Schwarzkopf. I stopped my maintenance alert folks and told them to expedite the refueling of the Reeves aircraft as our DV wanted to get going on his fishing trip.

As soon as I got back to the office, I asked my admin sergeant to find out who this general was. He came back a short time later and told me he was the commanding general of U.S. Central Command at MacDill Air Force Base in Tampa, Florida. It wasn't until January 1991, less than two years later, that I realized this unannounced visiting general was in fact the one and only General "Stormin'" Norman Schwarzkopf, the Commanding General of Operation Desert Storm during the liberation of Kuwait.

Mistaken Identity. I was able to take two weeks of very needed annual leave in late June and early July and had a great time with all of the family. Since I was returning from leave, I chose to fly commercial from Anchorage back to Galena and was travelling in civilian clothing. Markair had two commercial flights between Anchorage and Galena every day so the transportation support for our squadron was superb. While waiting for my flight I saw a young-looking person with an obvious military haircut waiting for the same flight that I would be on. I asked him if he was in the Air Force and if he was going to Galena for his remote tour. He told me he was, that he was a vehicle maintenance expert, and that he was a little apprehensive about going to Galena. He asked me if I knew anything about the commander at Galena and if the Commander was a good guy. I told him that the rumors I heard were that the Galena commander was a pretty good guy. I was very amused at his question but never let on that I was the Galena commander. We boarded our flight and arrived at Galena right on time. As we got off the airplane and were waiting for our baggage, I saw the young airman looking kind of confused about what to do next. My vice commander had just arrived at the air terminal in my staff vehicle and was ready to take me to my quarters. I stopped the young airman and told him to join me in my staff vehicle for a ride to the billeting office to get his dorm room assignment. What a sheepish look came over his face as he asked, "Are you the Commander?" I was very amused about this whole situation and learned a very valuable life lesson: never assume anything about strangers you may meet.

Annual Aerobics Run. The Air Force in 1989 had an annual requirement that every military member of the force was required to be tested in a 1.5 mile run to determine overall fitness. I did have a few of my older sergeants ask me if we could just waive the annual aerobic test since we were all serving on a remote unaccompanied tour. My answer was an unequivocal…no. Everyone would run the annual aerobics test. As an added incentive I challenged the entire squadron, all 325 military members, that anyone who could run the 1.5-mile test faster than the commander would earn a three-day pass. Because of the varied work schedules our admin folks set up three separate times to run the test on July 16. They had laid out the 1.5-mile course on top of the dike that circled the base. I thought I may have over-promised on the three-day pass and would have multiple members taking me up on my promise. The first run was at 0700 hours and there was probably one hundred squadron members at the run. Everyone was skeptical of the forty-two-year-old Commander who challenged them all. Fortunately, I had been running a good bit since the hassle of the flood evacuation, the UEI preparation, and the actual inspection. I knew I could run a fairly good race. In the 0700 race I managed to beat everyone who accepted the challenge and finished with a respectable eight minutes and twenty-two seconds. I'm not sure if the rest of the runners just let me beat them, but I was one third of the way through my wager. The next run was scheduled for 1130 hours. Again, there about one hundred runners for the second race. I had passed on lunch that day and still made a good time of 8:27 on this race; no one finished in front of me. The final run of the day was scheduled for 1600 hours. I ate a protein bar at about 1430 hours and felt pretty cocky about my wager with the squadron. There were about eighty squadron members at this third run. I was more than ready to shine my ass once again. The run was going just fine, I could see the finish line up ahead, and thought I had it in the bag when of my "Firedogs" from the fire department passed me like I was standing still. I was very surprised as this individual had been on one of the butt patrol teams. My time on the third run was 8:20, but I was beat by a good ten seconds and had no choice but to award the "Firedog" a three-day pass. That was the last time I challenged my young airmen to any athletic contest.

Operational Readiness Inspection (ORI) Rumors. The latest rumor that was going around Galena was a potential ORI in the very near future. It is highly unusual for a commander to get two major inspections (report cards) in the same calendar year. Remember, we just completed the UEI in mid-May, and I thought the inspection cycle was over for this year at Galena. The rational I heard about the ORI was that since Pacific Air Forces (PACAF) was now taking over the top command leadership of all Alaska activities, the 21st TFW and associated units would be subject to an ORI to determine a baseline combat capability before the change in force structure. PACAF headquarters is located in Hawaii and the new designation for the Alaska forces would be the 11th Air Force headquartered at Elmendorf Air Force Base.

11th Air Force would be tasked with the mission of Air Force planning, conducting, controlling, and coordinating all air operations for PACAF. This organization would also be the provider of all U.S. forces for the Alaskan Command and the North American Aerospace Defense Command Alaska region.

Alaska Command and 11th Air Force had a very long history in the overall U.S. defense structure of Alaska. 11th Air Force was established in 1941 and tasked with providing air defense for Alaska and engaging in combat operations primarily in the Aleutian Islands and northern part of the Pacific. Alaska Command was re-designated Alaskan Air Command (AAC) and provided air defense of Alaska during the entire Cold War. With the collapse of the Soviet Union, AAC was realigned under PACAF starting in 1989.

With all these structural re-alignments, I guess I could understand that PACAF needed to establish their war fighting capabilities base line. I will explain in greater detail the concept of the ORI and how our squadron did during this important inspection in a later chapter in this book.

Assignment News. Finally, I got confirmation of my follow-on assignment that would take place in October. My persistence with the general that I had worked for at Headquarters Alaskan NORAD Region a few years ago paid off, and I was advised I would be moving to Peterson Air Force Base in Colorado Springs, Colorado for an assignment at NORAD Headquarters as the chief of the NORAD Counter-Narcotics Planning Division. It wasn't the exact assignment I was hoping for, but it kept my career goal of never being assigned to the Pentagon intact. I pressed the personnel folks to get my orders to Peterson finalized as I wanted my dear wife to fly to Colorado Springs and find us a house so my three children could start school in August. She was staying with her cousin in Colorado Springs and had just completed a half day of looking at homes to rent. The timing was perfect. I got a call from a good friend of mine that was stationed with me in Alaska and he had just moved his family to Colorado Springs that month. As he was walking around his new neighborhood, he noticed a for rent sign being put up on another house in his neighborhood. The individual who found the rental house would be on the NORAD staff with me, so how could I turn him down? I immediately contacted my wife and gave her the address of the rental house. She was there in less than two hours and rented the house on the spot. All the pieces and parts of the new assignment seemed to be falling in place.

I just don't know what I would have done without my spectacular wife during my entire Air Force career. Earlier in my career she moved by herself from New Mexico to Washington state, bought a house and a car by herself while she was eight months pregnant with our third child, and was keeping track of two and four-year-old sons. Also, a month later, she gave birth to our daughter in the Fort Lewis hospital in Washington state without me. I was on Air Defense Alert in a F-106 in another state. Military wives are very special, and mine was the very best! In normal circumstances I would be entering the FIGMO (finally got my orders) state of mind and would be shutting down my activities in preparation to leave Galena. However, that certainly was not a status I could fall into as the commander of Galena as I still had two and a half months of intense leadership activities and challenges to face before I left Galena.

July Intercept Summary
    Zero intercepts

Chapter Thirteen
LEADERSHIP CHALLENGES
August 1989

Racial Issues. As described to me during the in-briefing from the former Galena commander there were some serious racial undertones that were festering at Galena. I had the senior master sergeant that I fired from the base supply function who had caused many racial issues by her activist leadership, numerous inappropriate racial comments by individual military and civilian members, members of both races in the squadron, and finally a closed door meeting with my first sergeant expressing a real concern about the racial atmosphere at Galena. I also had an end-of-tour out-briefing by a mid-grade sergeant who told me some service members were wearing Black Power t-shirts to the midnight meal for the past week. I attended a farewell party for six CE airmen that were finishing their tour at Galena, and I was the only white person out of thirty-five attendees.

The first sergeant and two other senior enlisted leaders (all three were black) at Galena came to me with a unique concern based on their observation of the people assigned to the squadron. While I really had not noticed the squadron makeup, they informed me that the mix between blacks and whites in the squadron had reached nearly 50 percent and about 75 percent of the senior level leadership in the squadron was black. This group of senior leaders who came to me was absolutely convinced that the Air Force Military Personnel Center (MPC) at Randolph Air Force Base in Texas was doing a social experiment at Galena. All three of these senior leaders who came to me had an average of almost eighteen years in the Air Force and had never seen such a percentage mix of blacks and whites in the same squadron. Four of my seven officers were black, but I really never gave the mix much thought. I was too busy trying to lead the squadron.

Since these senior leaders brought me this issue, I asked them for their recommendations to deal with this perceived racial mix problem. Their first comment was to contact the personnel office back at Elmendorf and see if they had any insight into the racial mix of the members of the squadron. Second, they requested that I keep the pressure up on those caught causing any racial issues, both black and white. And finally, they recommended that I address the entire squadron on this problem. Fortunately, as I said before, I had an excellent and very wise vice commander who just happened to be black and who advised me on how I should approach this critical issue. It was only appropriate that my vice commander went by "Ace" as his first name. Ace was without a doubt one of the most professional, dedicated, and competent officers I ever had the pleasure of working with during my entire twenty-six-year Air Force career. His recommendation on this critical racial issue was to deal with it head-on and tell the entire squadron that no racial insults or activity would be tolerated and that immediate corrective/judicial action will be taken. He also wanted me to impress on the squadron that we are all part of the team and we had a critical defense mission to accomplish. Each member of the squadron, regardless of their ethnic background, played a critical role in what Galena was all about. I asked the vice commander to schedule a commander's call for the following Friday. I also advised my officers that every member of the squadron must attend one of the two scheduled commander's calls. I really wanted to prepare for this critical event. As I put my thoughts together, I ran them by my vice commander for his input. Again, it was beneficial to have such a wise vice commander.

With the commander's call two days away, I called the wing commander at Elmendorf to give him a heads up on the racial issue we were dealing with at Galena, explained the social experiment theory that my senior enlisted leaders thought was taking place, and advised him that I would be addressing this issue with the entire squadron in two days. His only comment to me was, "Good luck…I know you can handle this situation." I also called a good friend of mine, the lieutenant colonel deputy chief of personnel at Elmendorf, and asked him if he could research the idea that MPC was doing a social experiment with the squadron at Galena.

The two commander's calls went smoothly, and I was very straightforward with the squadron members. I told them we didn't have time for racial divisions, that we all had a mission to do, and that each member in the squadron played a critical part in getting our mission accomplished. As recommended by my vice commander, I gave them a summary of some of the incidents we had dealt with in the past few months and the punishment I handed out. I think the information was well received as I didn't hear any grumbling from the airmen after the meeting and we did not have another racial incident while I was still at Galena.

Concrete Capers. One of the unique aspects of doing major construction work at Galena was that we were completely dependent on the status of the Yukon River. We could only get some of the large construction equipment on site, by river barge, during the time the river was open and flowing, usually between May and September. This was the case for some extensive concrete work that had been programmed for Galena for the past two years. Over three years, five new functional buildings at Galena had been constructed, but purchasing and installing the concrete for sidewalks had been put off for budget reasons. The paths to and from these new buildings were a muddy quagmire for the airmen trying to get from one building to the next. So, when the concrete plant arrived and docked at Galena, my civil engineering officer said that we needed to take advantage of river remaining open for at least the next four weeks. Once the river started to freeze the concrete plant must move back upriver before it got stuck in the river until the next ice breakup. The concrete plant operators agreed; however, we had already used up the money allocated for the current year for the concrete work that was desperately needed at Galena. I consulted with my civil engineering officer and my resource management officer and they said we could spend some of the money allocated to other base projects and take advantage of the concrete plant during the open river situation.

The weather was cooperating, in spite of summer rain showers, and we were making great progress on the new sidewalks. Late on the third day of the additional concrete pour I got an emergency call from the head of contracting at Elmendorf. He told me to immediately cease all concrete work until he and his staff could come out to Galena. Reluctantly, I ordered the concrete work to cease but to keep the concrete plant in port until this problem was all sorted out. The very next day the head of contracting and three of his staff, all in business suits, arrived at Galena during a heavy summer rainstorm. I took them to my office and heard their pitch that I had broken federal law by spending money that was not allocated for that fiscal year. I told them I was aware of that but felt mission requirements justified moving money from other projects to take advantage of the concrete plant that was able to remain in Galena. I got some very stern looks, so I told them to come with me and look at the concrete project.

I put them all in my command vehicle and parked just across the street from where the concrete work that was being done. Remember that it was still raining and the road and non-concrete paths were very muddy. I told them that once the squadron airmen left the post office on a daily basis and walked to the dorms or dining hall, they had to walk through the mud. I encouraged each one of them to follow me as I walked from the post office to the dining hall. They were none too happy with the walk as they all had mud flowing over their shiny dress shoes and up their pant legs. As we got to the dining hall for our scheduled lunch the conversation shifted to, "We didn't know the problem was this bad at Galena without these sidewalks." As lunch was finishing up and it was time to get them back to their aircraft for their return to Elmendorf, the head of contracting told me to continue to pour the concrete as long as the concrete plant remained docked in Galena. He also said that he would find the additional funds ($150,000) to pay for the over-budget concrete project. This turned out to be a huge win for the squadron as the airmen no longer had to trudge through the mud during their daily activities. As soon as the contracting experts departed Galena, I called the wing commander and told him I was not going to prison for the over-budget concrete work and that contracting had come up with the additional money to finish the entire sidewalk project.

Alcohol Dichotomy. A very weird set of events was presented to me in August. On August 25 I was notified that Galena had been recognized as the "Best in the Command" for alcohol sales in 1988. The Alaskan Air Command, Morale, Welfare and Recreation (MWR) Division sent a very nice plaque to Galena for our "superb sales efforts." I wasn't sure what I should make of this award because of the serious alcohol issues we had to deal with at Galena. Getting an award for the most alcohol sales in the command seemed ironic.

Two days later I got a call from the Alaskan Air Command social actions office that they would be sending a team to Galena next week to do a staff assistance visit (SAV) to determine just why Galena was selling and consuming so much alcohol. I immediately told my club manager to take down our command recognition plaque while the SAV was taking place. I am not sure if we ever got a good reading from the SAV team on the alcohol issue at Galena other than it was a tough one-year remote assignment, the airmen drank too much, and we should be more watchful of our squadron members and their drinking habits.

Drunks on Every Corner. In addition to the racial issues that were popping up and the SAV from the Social Actions office, we still had an alcohol abuse problem at Galena. As I was driving to the office at 0730 hours one morning, I noticed the security police parked near the front of the dining hall with all their lights flashing. As I got out of the vehicle, I saw that they were providing aid to a female sergeant that was lying on the ground at the base of the steps into the dining hall. She had shown up to her job as a cook at the dining hall very intoxicated and was sent back to her dorm, but she fell down the stairs in front of the dining hall. I directed the SPs to get a blood alcohol test (BAT) on her as she was obviously very intoxicated. Fortunately, our security police BAT tester had been repaired so we didn't have to send her to Galena City for the test. She blew a .223 BAT result – almost three times the legal limit. She had been partying all night in the dorm and was in no condition to report for duty. As I reviewed this sergeant's personnel record, we had sent her to the SABRE alcohol intervention program three months earlier. She did complete the program and was cleared to return to her position. Since she had a previous alcohol run in and was very intoxicated again, I had no choice but to present her with an Article 15 with a $200 dollar fine for two months and a reduction in rank by one stripe.

In another incident, I got a call at 0400 hours about a three-wheel recreational vehicle (RV) accident near the main gate. I did not respond to the accident but did direct the SPs to make sure they got a BAT on the driver if they suspected alcohol was involved. Sure enough, when I got to the office at 0700 hours the SPs advised me that the driver and the passenger were intoxicated and blew a .352 BAT, well above the normal .08 limit; they were fortunate that they didn't die from alcohol poisoning. These two individuals were in the maintenance section and both received an Article 15 and a two-week base restriction.

During the weekly officer's luncheon, I got a call that an individual had been found unconscious in the junior dorm. I walked over to the dorm and the responding security police team told me a local citizen who had somehow got on base and was drinking in the dorm. Since he was convulsing, the responding medics told me it looked like either a cocaine or alcohol overdose issue. Once he was stabilized, I had the SPs call the City of Galena police to get this person off the base. I also directed that this individual be restricted from the base for twelve months as he was not an employee.

Assignment Woes. I got a call in my dorm room at 0330 hours from a very irate women who said she was calling from Texas. It seemed her daughter, who was in my squadron, had gotten an assignment to a base in Florida. She didn't think that was fair because she wanted her daughter closer to her in Texas and wanted me to change the assignment. I knew this would be a very difficult call as the women was very vulgar and loud, screaming at me about the Air Force. She said that if I didn't immediately change her daughter's assignment to the Texas base, she would call her congressman and file a formal complaint against me. I wasn't really in any mood to explain the personnel assignment policy of the Air Force. I listened as long as I could and told her to please contact her congressman and see what he could do. I had dealt with many congressional inquiry threats but most of them never developed into a formal inquiry. The next morning, I had the young female sergeant come by my office and told her about her mother's phone call earlier in the morning. She said she was not surprised by her mother's call, but she definitely did not want to be stationed in Texas near her mother and was very happy with the assignment to Florida.

Operational Readiness Inspection (ORI). The PACAF/NORAD ORI team arrived at Galena on August 21 with an inspection team of thirty, two colonels, the head of the PACAF IG, and the head of the NORAD IG Team. We were in for a very detailed inspection. This would be the second "report card" of my command, and while I thought we were ready for the inspection, this is what we were facing:

1. ORIs are conducted to evaluate and measure the ability of units with a wartime, contingency, or force sustainment mission to perform their assigned operational mission.

2. Phase I will evaluate the unit's transition from peacetime readiness into a wartime posture.

3. Phase II will evaluate the unit's ability to meet wartime taskings.

4. Scenarios. IG teams should attempt to create a realistic environment for evaluation while ensuring safety is not compromised. ORI scenarios should evaluate both sustained performance and contingency response.

5. Major Graded Areas. Units with a wartime or contingency mission will be evaluated in four major areas: initial response, employment, mission support, and the ability to survive and operate (ATSO) in a hostile environment.

6. Performance. IG teams should focus on mission performance. Academic testing should not be used as a primary measure of readiness unless it serves a specific inspection objective, or hands-on performance cannot be observed.

7. ORI Frequency. Although optimum frequency for readiness inspections varies among MAJCOMs, an outer boundary of no more than 60 months between Operational Readiness Inspections is required.

8. ORI Grading. The five-tier rating system (consisting of the grades Outstanding, Excellent, Satisfactory, Marginal, and Unsatisfactory.

9. Common Core Readiness Criteria (CCRC). CCRC represent six basic, overarching readiness criteria that all MAJCOM IGs will apply to each area of their respective ORIs.

10. Threat. Is the unit able to implement and sustain appropriate measures to meet changing force protection conditions?

11. Safety. Does the unit safety program facilitate unit readiness?

12. Security. Were adequate security measures employed throughout the exercise? Were Operational Security (OPSEC) procedures incorporated into plans and followed throughout the exercise? Were proper Communication Security (COMSEC) materials available, as specified in tasked operations plans, to ensure mission accomplishment? Were COMSEC, Computer Security (COMPUSEC), and other measures employed to deny the enemy information?

13. Communications and Information. Were these operations effective?

14. Training. Were units properly trained and equipped to perform wartime duties?

15. Operational Risk Management (ORM). Were units applying ORM principles and concepts to assess the risks associated with their daily mission?

16. Contracted Functions. As a minimum, identify those contracts that are critical for the unit to accomplish its mission, sample the contractor's performance (through the Quality Assurance Personnel) as compared to the Performance Work Statement (PWS) requirements and evaluate the adequacy of PWS as written to satisfy mission requirements. Ensure the Quality Assurance Program and Performance Plan provide effective oversight of the contract and PWS.

17. Grading and Report. Contracted support activities grades will be five-tier and results integrated into the overall unit ORI report. MAJCOMs are encouraged to focus on the results of the contracted activity in support of the unit mission and to identify strengths and weaknesses.

The ORI was a very detailed inspection process that evaluated every aspect of our unit's war fighting mission in minute detail. The team arrived at 1800 hours on August 21, and the first task was to get the inspection team to their work locations and dorm rooms. We had practiced this reception step many times before the inspection as it is always a good thing to give the inspectors a good first impression. We were able to process the entire team in twenty-one minutes. They all had their rooms keys, Galena welcome packages, and directions to the base facilities. The only problem was the two inspectors who were designated to grade our in-processing procedures were at the club drinking as the processing steps were completed. I felt this would be a very long inspection since the IG team certainly did not give me a positive first impression.

I got a call in my office early the next morning from one of the security police IG inspectors about how we needed to get our security police to back off and not be so diligent on the security of the base. I immediately raised the bullshit flag and told him to send me the two IG colonels if he wanted me to back off the security of the base. Both colonels came by my office and started to chew my ass about my security police not letting their inspectors into certain areas of the base. I asked them if they really wanted me to reduce the security procedures during an ORI. We finally came to an impasse, and they said our security police would get an excellent rating if I relaxed the procedures a bit so some of the exercise scenarios could be run. I asked for the specific situations where their inspectors needed to get to certain areas on the base. I passed those exceptions on to my security police chief so his security forces would let the inspectors pass so they could simulate the necessary scenarios.

Day three of the ORI turned out to be interesting. I had opened the Galena Operation Center (GOC) as my battle staff dealt with many simulated wartime scenarios. We had many unexploded ordnance events, simulated personal injury events and many, many intelligence simulated exercise messages to deal with. We did a great job responding to each event when the next simulated inspection scenario was that the GOC was bombed and we had three simulated fatalities. The Defense Condition (DEFCON) had risen to a wartime level, and we very easily reconvened the GOC at the alternate location. We had set up backup communications equipment before the ORI, so we only experienced about seven minutes of down time while we evacuated and reestablished the GOC in the back up location. After about one hour in the backup GOC we were bombed again and had to move to our alternate, third-level GOC, my dorm room, where I had our communications experts install a secure phone in my room about a week before the inspection. The IG team was amazed that we had thought through this potential situation.

As the day went on, I was totally underwhelmed by the operations inspector. She was a senior airman (three stripes) that knew absolutely nothing about combat operations or how Galena worked. Some of her write-ups were absolutely ridiculous, and I was easily able to get them thrown out in the final report. For one of her write ups, she said that we didn't evacuate outside a 2,000-foot radius from the first bomb attack on our GOC. Here's the big problem with that: our entire base perimeter was less than 2,000 feet across. We would have had to evacuate to the middle of the Yukon River or to the marsh to the north of the base. As we saw in the UEI in May, the IG team was very unprofessional. We still had many more simulated events to deal with.

Next on the agenda was a practice scramble of our two alert jets. The maintenance and operations personnel performed flawlessly and both jets got airborne within their required five-minute launch window. While the F-15s were airborne we had a real-world in-flight emergency; a C-141 landed with hot breaks. Our maintenance crew and crash response team were right on top of the emergency and did all the correct procedures in response. While this emergency was not part of the inspection, the IG team was impressed with our response to these unplanned events.

Later that day we had a simulated Soviet missile attack on the base, and we all had to go to our shelters for about two hours. Most of the squadron response was good except for a few maintenance and supply knuckleheads who did not hustle to the shelter and were considered killed in action in the attack. These "lost" members of the squadron cost us a few points in the overall outcome of the inspection.

Days four and five of the inspection were uneventful, and I thought some of the inspectors were really getting into the minutia of our operations by reviewing our office files, regulations, and other mundane items. This took on the flavor an extended UEI, rather than testing our ability to conduct our wartime mission. The two IG colonels visited my office in the morning and wanted me to take them into the City of Galena so they could see just what an arctic Alaskan village looked like. When we entered the city store one of the colonels inquired about the most unique thing he could buy in Galena. I told him an eight-dollar gallon of milk, as the only way we got fresh produce or dairy products into Galena was by air delivery. Eight dollars was a lot of money in 1989.

Finally, at around 1500 hours on day five the IG team called an end to the ORI and released our squadron back to normal operations. We spent the rest of that day, until around 2100 hours, validating some of the items they documented in the write-ups. In the end I was able to delete three of the write-ups from the operations inspector who really had no clue about combat operations and, in the end, we got a good mix of Excellent and Satisfactory ratings for the ORI, with the overall rating being Satisfactory. That was OK by me as I thought the entire inspection team was less than professional. I would have given the IG team a Marginal rating. I attribute the lack of knowledge by the IG team to the fact that this was the first inspection done by the PACAF IG of any Alaska operations.

Off Limits. Many times, an Air Force base commander designates certain civilian establishments as off limits to military personnel. The reason for the off-limits designation can run the entire spectrum of reasons. In Galena, the one and only civilian bar – Hobo's – was designated off limits. A few months before I took command, one of our young airmen got in a fight at Hobo's and shot one of the locals. While all of the parties in the shooting were very drunk, the airman was arrested and charged with felony attempted murder. He was moved to Fairbanks to await a civilian trial. Even though this event took place before my command tenure, I was still concerned about a potential retaliation if my airmen went back to Hobo's. I made a command decision that Hobo's would remain off limits during my time at Galena. Every new person who arrived at Galena during my command was briefed that Hobo's was off limits to all military personnel.

Late on the evening of July 7 after I returned from my leave, I got a call at 2330 hours from a very drunk person who said that he was the owner of Hobo's and wanted to talk to the Galena commander. I told him I was the commander and asked how could I help him? He explained that my decision to make Hobo's off limits was killing his business and he may have to close his bar. I told him I understood his concern but that I also had a concern for the safety of my airmen and felt that there were still some local folks that wanted to even the score with our military folks for the previous shooting that occurred at Hobo's. I told him my decision was final but that I may review the decision after the trial of the accused airman was completed. I heard probably the most eloquent string of profanities I had ever heard before as he really exploded about my decision.

This wasn't the last of the off-limits issues with Hobo's. Later in July I got some "intel" from a senior sergeant in the barracks that a few of my airmen were in fact still going to Hobo's on Saturday nights. I called my security police chief to my office and told him I needed him to show up at Hobo's on the next Saturday night to see if the rumor about our airmen attending Hobo's was true. He said he would do that task himself and report back to me on the next Sunday morning. Early the next Sunday morning the senior SP reported back to me that he in fact had seen two airmen, other security policemen, at Hobo's enjoying the drinks and the local women. I told him to have both of them report to me early Monday morning. Both sergeants were waiting in my office on Monday morning and they came up with an incredible story that, on their own, they had heard from the rumor mill that Galena folks were still going to Hobo's and they wanted to catch them in the act. I called a BS on both of them and told them that I would begin processing Article 15 actions against them for disobeying my orders about Hobo's. I had already coordinated with Judge Advocate's Office at Elmendorf and got the initial approval to move forward with Article 15 punishments.

Then during the last month of my command I was notified that the owner of Hobo's had died from a heart attack and that the locals were so upset that they laid his body right on top of the bar at Hobo's for three days while they came up with a plan for a memorial service and burial. Only in Alaska.

Our New Soviet Friends. We were notified in mid-July that three Soviet aircraft would be passing through our airspace on August 6 to fly into Elmendorf and then on to Canada for an airshow. What a change in our response from the normal cat and mouse game of detecting and intercepting the many Soviet flights that flew near U.S. airspace. The mission of our alert F-15s was to intercept the three Soviet aircraft and then escort them to Elmendorf. The three aircraft were an An-225 cargo aircraft and two Mig-29 fighters. At the time, the An-225 was the largest aircraft in the world. It was 275 feet long, had a wingspan of 290 feet and a fully loaded weight capacity of 1,410, 958 pounds, and could fly at 430 knots. It was a huge airplane. The Mig-29 was one of the Soviet's front-line fighters that had just started to come into their inventory. It was a very capable and maneuverable fighter at fifty-six feet long with a wingspan of thirty-seven feet; it could fly at 50,000 feet at Mach 2.25. The Mig-29 was a comparable fighter to the F-15.

Since my operations officer was on leave for this event, I was the supervisor of flying for this mission and had been at the CAC since early morning to ensure all the support equipment was ready and the weather would be compatible for the intercept and recovery at Elmendorf. Although the weather was good, the regional operations center (ROCC) at Elmendorf was confused about the estimated departure time of the three Soviet aircraft. We were first advised of 0755 hours estimated departure time, then 0915 hours, and finally a "confirmed" time of 0945 hours. I wanted to make sure we did not screw up this intercept and escort mission as I knew the whole world would be watching us. At 0800 hours I coordinated with the Alaska NORAD Region fighter duty officer and directed the two F-15 pilots to go on battle station alert as we didn't have a real good feeling for the departure and intercept time of the Soviet aircraft. Battle stations meant the pilots would have completed all their preflight checklists, be strapped in their cockpits, and would be immediately ready to start their engines and taxi for takeoff. It's a good thing I did that because at 0840 hours we had an out-of-the-blue scramble order because the three Soviet aircraft showed up on the airborne AWACS radar. Our fighters were able to get airborne at the designated time, conduct a perfect intercept of the three aircraft, and escort them all the way into Elmendorf. I guess the Soviets were now our friends.

August Intercept Summary
August 6: 1 An-225, Cossack, 2 Mig-29s, Fulcrums

Chapter Fourteen
MORE LEADERSHIP CHALLENGES
September 1989

"**S**tolen" Commander's Vehicle. The normal resupply aircraft that came to the forward operating bases, Galena and King Salmon, usually arrived each Tuesday and Thursday bringing necessary equipment and supplies to both locations. Sometimes the aircraft arrived at Galena first, sometimes at King Salmon first depending on how the various delivery loads were packed. Our squadron had been anticipating the delivery of the new commander's vehicle, a 1989 Chevy Suburban. The current command vehicle was starting to show its wear and tear and was due for replacement. The Aerial Port cargo handlers notified me that the new vehicle would arrive at Galena on Tuesday, September 5. My transportation and communication folks were anxious to get the vehicle delivered so they could get the radios installed and then make the vehicle operational as soon as possible. My aerial port folks were waiting for the aircraft to arrive so they could get the vehicle off-loaded and turned over to the transportation/communication experts to start the upgrades. As the aircraft began off-loading, they noticed that the new vehicle was not on the aircraft. They immediately asked the load crew on the aircraft where the new command vehicle was. They said it had been off-loaded at King Salmon by the direction of the base commander there.

Their first action was to call me to the fight line and advise me of the "theft" of the new vehicle. My reaction was not to overreact to the theft. I asked my logistics folks to contact the King Salmon logistic folks and see if they could get the story on the commander's vehicle. The King Salmon aerial port staff told my folks that the vehicle was off-loaded at the direction of the King Salmon commander and that they were going to keep the vehicle for their own use at King Salmon. Well that was the wrong answer!

I next called the commander at King Salmon and asked him why he had "stolen" our command vehicle. He said his command vehicle was in pretty bad shape and that he needed the new vehicle more than we did as his transportation folks forgot to submit the paperwork for a new command vehicle the previous year. I told him that his decision was not the correct one and I expected him to return the vehicle on the very next contract flight from King Salmon to Galena. I next relayed my discussion with the King Salmon commander to my aerial port experts and told them to look for the command vehicle on the next contract flight from King Salmon.

After the next six flights from King Salmon to Galena did not produce the command vehicle, I advised the aerial port folks at Galena to report to me the next time a resupply aircraft from Elmendorf to Galena and King Salmon stopped first at Galena. That very next day I was advised the aircraft from Elmendorf was scheduled to arrive at Galena. I went down to the flight line and looked at the cargo manifest and noticed that a brand new, twenty-passenger Air Force bus was to be delivered to King Salmon following the stop at Galena. I directed my aerial port staff to off-load the bus and put it in the birchwood hangar for storage. I could hardly wait for the call from King Salmon.

Barely an hour after the aircraft arrived in King Salmon, I got a call from the commander at King Salmon that we needed to return his bus immediately. My answer was very direct, "We will hold your bus hostage until my command vehicle arrived back at Galena." After about three more weeks and numerous calls between King Salmon and Galena – and letters from Alaskan Air Command – our new commander's vehicle magically showed up at Galena and we returned the King Salmon bus the very same day. I am sure the AAC leadership thought the two forward operating base commanders were acting like children, which was probably true, but we got our commander's vehicle back.

Rape Trial. After nearly a two-month investigation, the judge advocate general (JAG) at Elmendorf decided that there was enough evidence to proceed with the rape trial that occurred in early July between one of my security force members and a member of our services section. There had been much discussion at Galena about what I would do with these rape allegations, but it was a serious enough that I felt I had no choice but to proceed with the court martial. The court martial was scheduled to begin on September 19 at Elmendorf, and I would need to bring the four key witnesses, the accuser, and the accused to the trial. I decided that I would also remain at Elmendorf to provide support not only to the witnesses but the accused and the victim as well, because the old adage about "innocent until proven guilty" was also part of the military justice system.

I had quite a challenge from the defense attorneys at Elmendorf as the month prior to the trial they requested my permission to allow the accused rapist time to take a ten-day leave before the court martial started. My answer was unequivocal…absolutely not, as two of the key witnesses in the court martial had already passed their one-year point in their assignment at Galena and were on administrative hold until the court marital was completed. It was not fair to keep these witnesses past their DEROS date while the accused was allowed to travel and be with his family. I held strong on this decision even after it was passed to the 21st TFW commander for resolution. Fortunately, the wing commander supported my decision.

Under the Uniform Code of Military Justice (UCMJ), a court martial is a criminal trial for members of the military who are accused of committing crimes listed in the punitive articles list of the UCMJ. Some of the crimes listed in this punitive section are similar to civilian crimes. Other crimes, such as desertion and insubordination, are specific to the military. In some cases, the civilian judicial system will coordinate with the military to try the non-military specific cases in civilian court. Many times, the civilian judicial system will defer to the military and let the military try the non-military specific crimes. In this particular case the civilian judicial system deferred to the military and a court martial was scheduled.

Because of the severity of the crime, and the fact that the defendant pleaded not guilty, a General Court Martial was prescribed. A General Court Martial has a military judge who oversees the trial, at least five members of the jury, and military appointed defense attorneys. The normal jury pool in a General Court Martial consists of commissioned officers. In this particular case, since the defendant was an enlisted member, he could choose to have at least two enlisted members be part of the jury pool. In this rape trail there were five commissioned officers and two enlisted members. So, the court martial was all set to begin on September 19 at Elmendorf Air Force Base. As soon as the court martial was convened I was told that I could not sit in on the deliberations as I knew four of the five officers on the board. Two had been former members of my flying squadron in Alaska and the other two were former staff members on the Alaskan Air Command staff who worked for me from 1985 to 1987. The court martial defense team requested that I be a witness if the defendant was found guilty because, as his commander, I may have some influence on the subsequent sentence.

Not being able to be inside the trial was no particular problem for me as I had many other things I could do while at Elmendorf. Finally, at 2145 hours on September 20 the verdict came in…not guilty on all charges. Talking with the jury members that I knew, they stated that the accuser in this particular case was just as culpable as the defendant because both were very drunk and, according to witnesses, she initiated the sexual contact and then had second thoughts on what had taken place. The defendant was obviously elated and called his wife back in Texas and told her the good news. I told the defendant that I would work on a plan to get him released from the remainder of his remote tour at Galena (two months early) but in the meantime ordered him to stay away from the accuser and any of the witnesses that testified against him. I also took the accuser aside and said that based on the verdict that she should move on with her life and ordered her to also stay away from the defendant. We all were able to return to Galena the next morning on the normal L-188 support flight, and I turned the not guilty defendant over to my first sergeant as his direct supervisor while we worked to get him an early release. Once back at Galena, I also ordered the female in this case to meet with her immediate supervisor and to stay away from the accused, just do her job, and not to drink so much in the future. All seemed well with this arrangement when I left Galena for Colorado Springs the next month.

Father-Daughter Talk. Since this rape trial was such a sensitive case for all the squadron members that were confined to Galena, I knew I needed to do something to get the word out about what actually had happened at the court martial. I was concerned that the rumor mill about the trial would cause havoc at Galena. I asked the two most senior female sergeants in the squadron to come by my office as I needed to discuss this rape trial with them. I felt that these two individuals had a sense for the feelings of our female squadron members. At the time we had fifty-six females and 287 males assigned to the squadron – a very potentially dangerous mixture exacerbated by the unaccompanied status of all the squadron members. These two female sergeants told me that the females were satisfied that I had brought the alleged rape to a court martial but were not happy with the verdict. I told the senior sergeants that I wanted to do a sit-down briefing with all the females and tell them exactly what transpired at the court martial and what each one of them should do in the future to prevent getting into the same situation. I had the sergeants review my talk outline and told them to coordinate with the first sergeant and set up the meeting as soon as possible. Was I ready to have a father-daughter talk to fifty-six female airmen? This was a real stretch for me as my own daughter was only eight at the time.

It was a very tense time as I had all the females assigned to Galena in the same room at the same time. We used the commander's conference room with the conference table moved to the side and fifty-five chairs set up facing the front of the room. I chose to sit in a chair at the front of the room and be as straight forward with the information and guidance as I could be.

I told them, "I am sure you have heard numerous rumors about the recent rape trial and that I wanted to give you the straight unvarnished answer on what took place at the court martial."

First, I told them what the verdict was…not guilty on all charges. Second, I relayed what two of my former squadron members that were on the jury said about the outcome of the verdict as I was not allowed to sit in on the trial since I knew four of the jurors. And finally, I gave them what I thought was some good fatherly advice. I had a range of female airmen from eighteen years old to the mid-forties in attendance. It was very tense when I told them that the not guilty verdict was delivered because there was much alcohol involved and some not too subtle messaging from the "victim" about her intent. The "victim" of this alleged rape was not in attendance as my senior female sergeants wanted to spare her from embarrassment.

I next told them that even though there were around 285 males on Galena away from their wives or girlfriends, they should be very assertive in their refusal to have sex. "If you don't want to get laid please watch your alcohol consumption and your own actions. Sometimes males have a hard time reading the subtle clues that you don't want to fool around."

I also wanted to make sure that none of my guidance would stop them from reporting any sexual action that they thought was inappropriate. I continued,

"Even though there was not a conviction in this case, my preliminary investigation told me there was sufficient evidence to move forward with the case and I would do it again with the same evidence."

I asked the women if there were any questions and the only question I got was if I was going to give the same talk to all the males on Galena? My answer was yes, and I did give a similar talk to the males at Galena in three separate settings.

In addition, I thought I should also give my talk to the "victim" in this case and had her report to my office with my first sergeant and I gave her the very same talk that I gave to the large group of females. I also reiterated my order that she stay away from the other individual in the case and to just do her job and finish her assignment with no further problems.

However, there was a follow on to this particular case. After I had left Galena and was in my new position at NORAD Headquarters, I got a call from the judge advocate office in Alaska asking my opinion on the character of the female "victim" involved in this rape case. It seems that a week after I left Galena this female, a cook in the dining hall, beat the snot out of the defendant when he was coming through the food line one morning. She was charged with assault and was being court-martialed for that offense. What a tragic turn of events. I never did get the final outcome of her court martial.

One final observation on this incident: After my father-daughter presentation to the entire squadron I was very sensitive about all things male and female at Galena. My usual procedures when I attended the monthly Galena City Council meeting was to take one of my senior enlisted airmen with me to give them some insight on how the city of Galena operated. On the very night of my final talk to the squadron about the rape case I was scheduled to bring one of the senior female enlisted members to the city council meeting. Being very sensitive to seeing me drive off base with a female in my command vehicle I made sure that I announced, over the police radio in my vehicle, that I was going into town for the Galena City Council meeting and that I had Senior Master Sergeant "Jones" with me for the meeting. My chief of security police kidded me about the radio call the next morning but fully understood what I was doing.

Best in the Air Force. I was notified that the Hennessy Trophy Competition evaluation team would be coming to Galena the first week of September. The Hennessy Trophy is an annual award presented to the best food service programs from across the Air Force. Winners are selected based on their display and execution of excellence in management effectiveness, force readiness support, food quality and production, employee and customer relations, training, and safety awareness. This recognition was quite an honor for one of the remote sites to even be considered for the evaluation as most of the food sources that we consumed at Galena had to be ordered one year in advance and barged into Galena when the river was ice free. The rest of the food items were flown in on a weekly basis as long as the harsh Alaska weather permitted the aircraft to arrive with the food. Our food services folks had worked extremely hard for the previous three weeks getting the facilities in top shape and ready for the evaluation. We knew the competition would be difficult, and we had many obstacles to overcome to compete with other world-wide Air Force dining facilities. During the last week in September we found out that although Galena did not win the overall Hennessy competition, our food service superintendent, a master sergeant, had won the prestigious Hennessey Travelers Association award for excellence, identifying him as one of the best food services professionals in the Air Force. This food services expert was the very same sergeant who worked so hard to feed us during the flood evacuation in May. Receiving this award was quite an accomplishment for our small base and a real honor for our very own food superintendent.

Hunting and Fishing in Alaska. Alaska is known for its world-class hunting and fishing opportunities. People from around the world pay thousands of dollars to fly to Alaska and take advantage of the abundant wildlife and fish. Galena had a very robust Morale, Welfare, and Recreation (MWR) support unit which rented out six 18-foot Tracker boats with sixty-five-horsepower Evinrude outboard motors that could easily handle the swift currents of the Yukon River. So fishing was a very popular activity for the squadron during the months that the river was free of ice (late May to late September). I had fished many times where I grew up in Oregon and had done many fly-in fishing trips while stationed at Elmendorf.

One issue that I had to use my leadership skills was for the issuance of Alaska fishing and hunting licenses to our squadron members. A local store in Galena City that sold fishing and hunting licensees told our squadron members that they had to buy nonresident licenses. The cost for a nonresident license was nearly triple the cost of a resident license. I called the local Alaskan fish and wildlife officer in town and told him that I didn't think it was fair that members of the squadron, who all had an Alaska address while living at Galena, should have to pay nonresident fishing and hunting license fees. He told me he would get back to me. Well, low and behold, he called the very next day and said as long as a member of the squadron had a mailbox address on the base at Galena they could purchase their licenses at the resident rate. That was a huge morale builder for our service members.

So, I linked up with two of my experienced senior enlisted members and our chaplain to enjoy Alaska's bounty on the river. The current in the main Yukon River was very strong, so we worked our way into a few of the many sloughs. The fishing was incredible. Since I was the commander, I made an edict that, on my boat, you could not keep a fish if it was less than thirty inches long. That was really no problem as it seemed that on each cast, we were able to catch an incredibly tough fighting fish, the mighty pike. We caught our limit in less than three hours and headed back to the base with our huge haul of fish.

I did have quite a few squadron members who wanted to take advantage of the superior hunting opportunities in Alaska. I was able to ensure our squadron members could also get resident hunting licenses for moose and caribou. We were able to give each hunting party from the squadron a set of handheld radios to use in case they got into any trouble during their hunts. One of our senior civil engineering civilian employees had his own aircraft, a Cessna C-180, parked at the Galena airport. I asked him if he would be interested in letting me fly with him to do a "safety check" of the numerous squadron hunting parties that were in the field. He was more than happy to oblige. We knew the general locations for each of the hunting parties and flew over the area to make radio contact with them. They were all doing OK and a couple of the teams had already bagged their animal. There were a couple of requests from the hunting parties asking me if I had any visual contact with animals in their area. That was a very difficult question for me as I saw numerous animals very close to the hunting parties but could not tell them where the animals were as it is unlawful in Alaska to doing any aerial spotting of animals during the hunting season. Doing a safety check on each team was enough "leadership" the hunting parties needed from me. Getting to fly again was also a reward for me as it had been over a year since I had flown an airplane.

"Dead" Clock Radio. We had a very funny situation occur in mid-September when our squadron chaplain went on a one-week leave back to the States. He forgot to turn off his clock radio when he left, and the alarm blared loudly for most of the first morning he was gone. I asked my services officer if he could get the key to the chaplain's room and turn off the clock radio alarm. My operations officer had a better plan: he had just purchased a new Magnum,

.50 caliber Desert Eagle handgun and thought we should take the clock radio to the Galena base dump and put it out of its misery. All the officers piled in my command vehicle, drove to the dump, and proceeded to "kill" the clock radio with the very powerful handgun. Obviously, there was not too much left of the clock radio but we picked up all the pieces and brought them back to the chaplain's room and put them on the nightstand with a note that we had to "kill" the alarm because it was keeping us all awake.

September Intercept Summary
   September 13: 1 Tu-95 Bear H

Chapter Fifteen
TIME TO GO
October 1989

Drug "Bust". One of the positive results of my "leadership by wandering around" concept paid dividends in many ways. I felt that the folks in my squadron were comfortable bringing me real issues to deal with. Remember the fake OSI agent story that was brought to me by a young airman? I knew that the next piece of information needed to be dealt with immediately as there was a rumored drug problem within the squadron. I had only two weeks left in command at Galena and didn't want to leave this problem to my replacement. While I was at one of the work sections, a senior sergeant told me that he had heard rumors that there was a drug problem in one of the dorms. He said the rumor was that a civilian employee in the civil engineering section was the main drug lord for this issue. As soon as I got back to my office, I contacted my chief of security police so we could come up with a plan to deal with this drug problem. I addition, I contacted the FBI office in Anchorage to get some advice on how to proceed. The FBI agent asked if we had access to any Air Force drug dogs at Galena so we could first find the drugs and then determine who the dealers were.

The security police squadron commander at Elmendorf had been one of my students when I was a Faculty Instructor (FI) at Air Command and Staff College at Maxwell Air Force Base between 1983 and 1985, and we had a good relationship. The security police commander said he would facilitate getting the drug dogs to Galena. We just needed to coordinate a date and time, in addition to having a pretty good idea of where the drugs may be in the dorm. He also advised that any information about the drug sniffing dog mission be closely held or we may tip off the druggies that we were coming for their wares.

While I did not share the fact that I was going to bring the drug dogs to Galena, my chief of security police advised me that the Galena City police were now also aware of the possible drug activity on Galena and they had started an investigation into where the drugs were coming from and how they are getting in to the Galena system. They had also noticed an increase in drug related activity in the city. The Anchorage FBI office called and said they had a lead on the drug movement from Anchorage to Galena. Their sources said that the drugs were being shipped to Galena on the Reeves Air weekly L-188 resupply aircraft. They had a plan that our local military police, with the help of the drug dogs, could meet the arriving aircraft at Galena and try to intercept the drugs before the drugs got into the system.

It was now time to get the drug dogs to Galena to assist in the inspection of the arriving aircraft and to sweep the suspected dorm for drugs. I contacted the security police squadron commander at Elmendorf and requested the dogs be at Galena mid-morning on October 7. Since I didn't want any surprises, I had to advise my operations officer that the drug dogs would be arriving the next morning on a C-12. My operations officer said he would have his folks do a special handling procedure on the arriving C-12 and park the aircraft out of site of the main part of the base. Since I wanted this operation to be completely under cover, I didn't advise my chief of security police of my plans until one hour before the dogs' scheduled arrival. He would have plenty of time to get his inspection team together. I had requested the dogs arrive at least thirty minutes before the scheduled Reeves Air L-188 arrival.

I was certainly anxious to get this drug situation sorted out and stop the flow of the drugs through the squadron and then into the city of Galena. I was all prepared to meet the C-12 arrival with the dogs but was caught off guard by a call from the 21st TFW commander about the upcoming change of command ceremony. Right during the call with the wing commander, I got a very excited call on my hand-held radio that I needed to get to base operations as soon as possible as there was a major problem.

What could be going on? As soon as I arrived at base operations, I was met by my operations officer who told me there was a problem with the two drug dogs. It seemed that both drug dogs had become very air sick during the flight and were vomiting all over the ramp. I discussed the dogs' conditions with the senior dog handler and he said the dogs could not do their work until they recovered from the air sickness: about a forty-eight-hour delay. So, the fact that we now had drug dogs on base was certainly passed around the entire base and our element of surprise was shot. There would be no drug dog sweeps taking place today.

Just as I was dealing with the puking drug dogs, I got a notification that the Reeves Air flight, maybe with the drugs on board, scheduled to arrive in thirty minutes, had been diverted to King Salmon and would not be arriving at Galena until tomorrow. Our well-planned out drug bust was an utter failure, and we never had the opportunity to find the drug dealer or drug users during the remainder of my time as commander. I believe the fact that we had the drug dogs on Galena to search for drugs may have actually stopped the drug activity that was rumored to be going on. I never got an update from the new commander about what finally took place with drug operation.

Forged Paperwork. About one week before my change of command and departure from Galena I got a call from a munition's maintenance squadron commander at Nellis Air Force Base, Nevada about a former master sergeant who had been assigned to our squadron and had recently transferred to Nellis. The sergeant had complained to his commander at Nellis that he was not awarded a medal on his departure from Galena. This commander at Nellis was requesting that I reconsider giving a meritorious service medal to the master sergeant. I could barely keep my composure as I found out that this master sergeant, who was in charge of all our weapons storage facilities, had falsified his quarterly inspection of our Aim-9 missile inventory and had not completed other key inspections while he was in charge. I said I would absolutely not approve a medal for this individual and I was inclined to re-write his efficiency report because of his negligence. The conversation with the commander came to an immediate halt and he said, "I understand, just forget my request for a medal."

Farewells. During my last two weeks in command I wanted to visit every work section on the base and thank each member of the squadron for all their hard work and dedication to the mission of our squadron. It was an amazing experience as I there were so many new faces in the squadron. I felt like a new member of the squadron. I was very proud of the way the base looked and even though I was leaving I felt great pride in the way the base had improved in so many areas. At almost every stop, each shop had a going away gift for me. I felt very proud to have been a member of such a close-knit team during the previous twelve months at Galena.

As I was sitting in my office two days before the change of command finishing up some paperwork, I got a call on the command radio that there was a fire in the club. Oh great, where would we have the change of command reception if the club was unusable? As I arrived at the club, I saw two fire trucks parked outside with their lights flashing but no hoses going into the building. What was going on? As I entered the main ballroom of the club there were about 200 of the squadron members "waiting" for the boss. I got a standing ovation from the attendees and then was presented with more going away gifts, including a U.S. flag that had flown over the base during my command, numerous squadron plaques, and other memorabilia. I was more than pleased and thanked each member of the squadron for their dedication to the mission and that I expected the same support for the new commander.

Replacement Arrives. My replacement arrived at Galena on October 10, two days before the formal change of command. On that day I was reminded that even though I was the commander someone else occasionally determined my schedule. My replacement was due to arrive at Galena at 1000 hours. Just as I was driving to base operations to greet my replacement and welcome him to Galena, I got a call that another C-12 was also just landing with a code on board. Oh great, just what I needed. I drove over to the second C-12 and found that Alaskan Air Command (AAC) vice commander was on board and wanted a tour of the base. I asked my vice commander to pick up my replacement and get him started with his normal in-processing while I escorted the AAC vice commander. Fortunately, the visit with the AAC vice commander was short as he was also on his way to King Salmon. I never found out why we were not notified of the visit.

Once I was finished with the AAC vice commander, I joined all my officers and my replacement at lunch in the dining hall. My replacement had many questions, and I did the best I could to answer all of them. I must be honest, I was a little concerned by some of the questions he asked, and I hoped that he would not change some of the programs I had worked so hard to put in place… but what the heck, that would not be my problem any more in two days.

After lunch my replacement and I sat down to cover the key issues that he would be dealing with in the very near future including the delayed drug bust, the simmering racial issues, and the move into a new dormitory for the officers in the squadron. I really had mixed feelings about leaving Galena as I had put my heart and soul into this squadron for a year.

Change of Command. The formal change of command ceremony was scheduled for October 12 in the base gymnasium. The turnout was great as four C-12s flew into Galena carrying the wing commander and a number of my personal friends that I had at Elmendorf. Almost 250 members from the squadron, the Galena mayor, and the entire Galena city council were also in attendance. As is the custom, I was given the opportunity to speak first and I went down a list thanking all the professional people in the squadron and wing who had made this command assignment so memorable. The club had put together a first-class reception, but I only stayed for a very short time as this reception was for the new commander and the protocol was for the "old" commander to just fade away.

I had my admin sergeant take me to base operations and I boarded the C-12 waiting to take the dignitaries back to Elmendorf. After about one hour the dignitaries arrived at the flight line, and we taxied out for takeoff. As we taxied by the base fire department the entire fire department staff was standing at attention in front to the building and they had two large fire trucks spraying water over the C-12 as we taxied out. Once we got airborne the pilot called back to the passengers and told us he had a landing gear problem and needed to return to Galena. As he rolled out on final approach for landing, the wing commander started to laugh and told me this was just a "fake" emergency and wanted me to think I was not going to get out of Galena today. Really funny! But as we did a low approach, I saw many members of the squadron lined up on the ramp waving good-bye to the boss.

Family Escort. My older brother, a former Air Force officer, was scheduled to be in Anchorage on business the day I left Galena, and he offered to meet me at Elmendorf with a rental car to take me to the Anchorage airport; he and I were scheduled for the same flight into Seattle. As we drove down to the Anchorage airport, he told me that he had been able to upgrade our seats to first class. Wow – my first time flying in first class, I could get used to that.

Once in Seattle we contacted the car shipment company that had shipped my Honda Accord from Anchorage to a location near the Seattle airport. We picked up the car, and I was on my way to a new adventure.

October Intercept Summary
    October 7: 2 Tu-95 Bear Hs
    October 10: 2 Tu-95 Bear Hs

Chapter Sixteen
THOUGHTS ON LEADERSHIP

What Leadership Means to Me. My career up to this point, at the twenty-year mark, taught me many lessons about leadership and what the leadership equation means to me. My time as the commander of the 5072nd Combat Support Squadron at Galena certainly stretched my leadership skills in so many areas and gave me the opportunity to reflect on and further develop my own tenants of leadership. I thought I would summarize my views on leadership and explain how I developed some of my tenants of leadership. The skills and lessons learned during my time flying combat missions in the war in Southeast Asia in the early seventies certainly played a huge foundational role in the development of my leadership style. I would not trade my wartime experiences for anything because those experiences made me the person I am today…good or bad, this is who I am. I learned to do the right thing even if it was not politically correct at the time. I also learned to do it now…not to procrastinate with decisions. If I received better information after I made the decision, I could change the decision, but I never ever waited for perfect information on which to make a decision. This wartime development of leadership skills directly applied to my time as the commander of Galena. These growing leadership skills came into play during almost every decision, large and small, that I had to make at Galena. This experience certainly helped form my own personal concept of Air Force leadership. Leadership to me is simply the art of influencing and directing your people to accomplish the assigned mission. So, if you will indulge me a bit, here are my key tenants of leadership:

Mission. This is absolutely the most important factor in my leadership philosophy. We are all assigned to do a specific mission on each one of our Air Force assignments. In my career this included flying combat, testing air to air missiles, investigating aircraft accidents, sitting air defense alert, leading an Air Force maintenance unit, teaching at our Command and Staff College, being responsible for the readiness of an entire command, commanding a tactical flying squadron, and being a base commander in Alaska. For each mission I had an important role to play in the overall defense of our country. Some were certainly more exciting and sexier than others, but each mission I was given defined my being. It was the focus of all my activities each and every day. As long as I stayed focused on my assigned mission, I was earning my pay.

Former Air Force Chief of Staff, General Curtis E. LeMay stated, "No matter how well you apply the art of leadership, no matter how strong your unit, or how high the morale of your men and women, if your leadership is not directed completely toward the mission, your leadership has failed."

As a leader in each one of these missions I felt that it was my role to instill the concept of "mission first" to everyone I came in contact with. This was especially true during my two command assignments, as the commander is always expected to guide the entire direction and energy of the squadron toward mission accomplishment. At Galena, the mission was having two fighters on alert twenty-four hours a day, seven days a week and providing the necessary support to do that mission. Anything else was not important.

People. This key element of leadership is almost, but not quit as important, as mission. Obviously, the people of any unit are the ones who carry out the mission, and it is critical that each member of the unit know exactly what their role is to accomplish the mission. It does not matter if it is the pilot of the alert fighter or the squadron admin sergeant, each has a critical role in carrying out the mission. When I was the commander of the tactical flying unit, I tasked every member of the squadron with the responsibility to raise the BS flag if there was ever a situation developing that looked unsafe. This directive applied to the pilots and all the way down to the admin clerks and life support personnel in the squadron. Every person has a critical role to play and every action they take supports the assigned mission. One of a leader's key responsibilities is the care and support of the unit's personnel. Successful leaders continually ensure the needs of their subordinates are met promptly and properly. Take care of the people, and they will take care of the mission.

Communication. This leadership trait couldn't be clearer. The leader must communicate to all members of the unit from the top echelon to the very lowest ranked person. Each individual must know their role and how their role, however minor, plays in the overall accomplishment of the unit mission. A leader must constantly communicate to the members of the unit what is important as it relates to the mission. At Galena I was always talking to the support folks, for example the cooks in the dining hall, telling them the important role they played in ensuring the pilots on alert had the very best food they could make them so that the pilots were mentally sharp and ready to fly the intercept mission. I made it a point to visit each and every functional shop on Galena at least once a month to express my thanks and remind them of the important role each played in supporting our mission.

A corollary part of the communication piece of leadership is for the leader to listen to his people. I had four good examples of this process at Galena. First was my normal routine of sitting with random airmen at the dining hall to get their perspective on what was going on in the squadron. Second, I instituted the Commander's Gram Program where any member of the squadron could communicate directly to me on any issue that they thought was important. Third, I held monthly commander's calls where I would pass important information and field questions from squadron members about any subject they wanted. And finally, I made an effort to get to each functional area at least once a month to again get reverse communication on any issue that was of importance to the members. I felt this two-way communication process really gave me a feel for where the squadron stood on any issue and gave me the necessary insight to address a minor problem before it became a major one and impacted the performance of our mission.

Standards. Any organization has numerous standards that must be followed to ensure the mission is accomplished successfully. Part of the standards factor of leadership is the need to continually communicate those standards to the entire unit and consistently enforce the standards across the board with no exceptions. The technique I used at Galena to communicate the established standards to every single person who arrived at Galena was to present an in-briefing, directly from me in my office, within the first twenty-four hours of their arrival at

Galena. I did break the briefing down into two groups: officers and the top three enlisted ranks, and then everyone else. I also conducted mandatory end-of-tour (EOT) out-briefings with everyone leaving Galena to get a feel of how the squadron was doing with the mission and the established standards.

After communicating the standards, leaders must consistently enforce them. One slip of letting the folks not follow a standard is opening the door to more noncompliance with the standards that have been set. A good example at Galena was when we lost our base barber. I had a few folks ask me if we could just waive the haircut standards until we got a new barber. My answer was no way…and that led to numerous barracks barbers trying their hardest to ensure each member was in compliance with the Air For grooming standards. There were a few very bad haircuts but everyone made do and held to the standards.

Lead by Example. It is nothing short of complete failure of a leader if the members of a unit feel that there are things that the unit members must do but not the commander. I learned this concept flying fighters in the Air Force. Whenever we flew in a multi-aircraft formation we always looked to the leader for guidance. We would follow his lead on every maneuver even into the ground if that is where the flight lead would take us. So just watching what our flight lead did on every flight made us want to do the same thing when we were in the lead. Three examples of leading by example at Galena were my going to the front of the line for the annual flu shot, leading the squadron on all three annual aerobics runs, and pitching in to pick up trash around the base after the spring thaw. I would never expect the squadron members to do things that I would not do.

Decision Making. I learned this trait from my flying as a forward air controller (FAC) in southeast Asia. As a FAC I had to make life or death decisions on the spot on almost every mission and then stand by those decisions. FACs didn't always have perfect information to make a decision, so we had to act on the information we had at the time. In peacetime the decision process was no different. I don't believe I ever made a perfect decision at Galena because I don't believe I ever had perfect information. At Galena I told each one of my senior leaders that I would always make a decision based on the information presented to me. I also reiterated that I had no problem changing my decision if they came to me later with better information. You just can't wait for perfect information or a critical decision will never be made. A leader must have the self-confidence to make timely decisions. The leader must then effectively communicate the decision to the unit. Decisiveness includes the willingness to accept responsibility. Leaders are always accountable – when things go right and when things go wrong. An example of this was when I ordered the concrete folks to continue to pour much needed concrete on the base even though I knew we were being "creative" in paying for this additional concrete. I was ready to live with whatever the outcome would be as I knew this additional concrete was for the good of the squadron and our mission.

Teamwork. One skill that a leader should always encourage and foster is the concept of teamwork. Any mission assigned to a unit has many parts that make up the whole. Intercepting Soviet aircraft out of Galena was not just the fighter pilots flying the intercept mission. There were so many parts behind the scene that were critical to the final result of the mission. If any one of those parts was missing the mission would fail. So, the leader's role in this trait is to define, encourage, and reward each part of the team supporting the overall unit mission, no matter how minor the task appears. In support of our F-15s at Galena, we had aircraft maintenance experts, fuel experts, munitions experts, air traffic control experts, housing experts, food services experts, supply experts, and administration experts. The list goes on and on. My role, then, as the leader was to define why it was so important that, for example, the fuel experts were outside in -60°F weather refueling the aircraft so the aircraft could immediately be back on alert status for the next intercept mission.

When I was an aircraft maintenance officer earlier in my career, I was in charge of a 255-person back shop that repaired broken avionic parts. The back shop did not work on the flight line but instead repaired critical aircraft parts in a warm and environmentally-controlled shop. Fortunately, I was also still actively flying the F-106 and, to impress upon my maintenance team the importance of the work the back shop repair folks were doing, I would insist that if I ever ground-aborted a flight I would call the individual repair person in charge of that part out to the flight line. I was able to show them first-hand why their repair was so important, and when they didn't repair the part properly we lost an important sortie. The concept of teamwork is so critical to any organization. In the last part of this chapter on leadership there is a copy of guest editorial that I wrote for the Elmendorf Base Newspaper praising the role teamwork played in the success of the 5072nd Combat Support Squadron.

Integrity. A good leader must have total commitment to the highest personal and professional standards. A leader must be honest and fair. Integrity means establishing a set of values and adhering to those values. Former Air Force Chief of Staff General Charles Gabriel said, "Integrity is the fundamental premise of military service in a free society. Without integrity, the moral pillars of our military strength – public trust and self-respect – are lost."

While I was at Galena, I was tempted many times to go to the club and drink with the boys. That was something I had done more than I want to admit when I was flying combat in Vietnam. But deep down I knew that I needed to lead by example and not let myself be in a position to be questioned about my behavior. I used that same thought process by being sensitive to the male-female issue in the squadron when I would take a senior female enlisted person into town for the city council meetings and announce on the radio the name of the person I was traveling with.

Loyalty. This important three-dimensional trait includes faithfulness to superiors, peers, and subordinates. Leaders must first display an unquestionable sense of loyalty before they can expect members of their unit to be loyal. I always knew I had a boss even though I was the commander of the squadron at Galena and my direct boss was 350 miles away at Elmendorf. I always knew that his decisions were probably based on additional information that I did not have. The second key part of loyalty was to the people, the officers and senior enlisted personnel, who worked directly for me. While it was completely obvious that I was the boss, I wanted to make sure that each officer and senior enlisted member in my command realized the important role they played in our mission. I felt I had a responsibility to each one of them to mentor them as they progressed through their careers, just as key leaders had mentored me through my early career. And finally, I needed to be loyal to my subordinates or my junior enlisted force. I always have said that the enlisted force is the backbone of the Air Force and they do all the dirty work of every mission. To show loyalty to them I needed to set and communicate the key standards of performance expected of our mission and then carefully and evenly enforce those standards. Another element of loyalty was to recognize superior performance when observed, and I tried as hard as I could to prepare top notch efficiency reports and appropriate awards when they performed well.

Guest Editorial. While this assignment as the commander of Galena was a very challenging time in my career, I felt that many of the leadership abilities that I learned over the years were utilized during the year on the river. I was asked to contribute a guest editorial for the newspaper at Elmendorf Air Force Base about my command at Galena. That editorial is below and finishes out my memories of this very challenging and rewarding time in the Air Force.

"As I reflect on my last 10 months as Commander of Galena Airport, the time frame from 13 January to 25 May stands out as the most intriguing, exciting, rewarding and memorable. It was during this time I witnessed the most graphic displays of teamwork ever demonstrated during a peacetime, real world period. It seemed like the men and women of the 5072nd Combat Support Squadron (CSS) thrived on the challenges they encountered; the more they were challenged, the better they performed.

"The period began when somebody turned the thermostat off. Temperatures plunged to an average of minus 60 degrees for 18 straight days, peaking at minus 70 degrees. Pipes froze or burst, vehicles had to be moved inside to start, two out of three power generators went down and the third was on the way down because of potential overload. Our dinning facility was under renovation and food had to be prepared at one location then transported to another for serving. Portable heaters had to be used to prevent the loss of our last generator and our heating fuel supply had reached a critical low point.

"Through it all we survived because of teamwork. We survived and the intercept mission of Galena remained on alert status the entire time. Our plumbers fixed pipe after pipe, the fuels people coordinated with Elmendorf to get a 700,000-gallon emergency air delivery of fuel, then off loaded it in minus 60-degree weather. While all of this was happening, the men and women of the 5072nd CSS made time to collect firewood and clothing for several elderly families in the local community. Each person in the squadron had to do their part for things to work out. It was a fine-tuned team that brought us through the coldest winter in Galena history.

"Shortly after the cold spell came the breakup of the ice on the frozen Yukon River. The mighty Yukon River with currents in excess of seven knots has the potential to devastate everything in its path when not confined to its natural path. We developed, reviewed and finally practiced our plan for an emergency flood evacuation. We were planning another exercise with our flood evacuation plan when Alaska State officials notified us the Yukon had risen to 140 feet above normal river level and was expected to top out at

154 feet within hours. Colonel H.S. Storer, 21st Wing Commander gave the word to "evacuate your people." Immediately base operations coordinated the launch of 6 C-130 Hercules aircraft from Elmendorf.

"The 51 people left behind at Galena began flood mitigation preparations. Complete activities were dismantled, computers from the first floor were relocated in offices and hallways upstairs; unneeded vehicles were moved to higher ground and boats were moved to key locations for emergency, last minute getaways. The Galena Operations Center began 24-hour operations. Fuels people became cops, food service people became projectionists, and civil engineers became administrators. While monitoring the river's activities, coordination with Elmendorf revealed the support to our evacuated members to be flawless. With very little advanced notice, Elmendorf provided aircraft and ground transportation for an unexpected 270 people. A section of Elmendorf's family housing area was instantly converted into temporary quarters.

"After nine days at Elmendorf the "all clear" signal was given and the men and women of the 5072nd CSS began the ordeal of returning to Galena and putting their lives and their mission back in order. But the 5072nd people were returning only seven days before the Alaskan Air Command Unit Effectiveness Inspection (UEI) team was scheduled to inspect our squadron. The final week before the UEI we had to unpack complete functions, reassemble flight line equipment and activities, relocate and reconnect computers. On top of this we had to complete work on the UEI work center, organize billeting and transportation for the inspection team and refine everyday procedures to meet the very in-depth review Galena would receive. Within five days after returning from Elmendorf, things fell into place. Equipment was back on line, final tests of radar and flight line equipment were complete. Meanwhile our alert aircraft had returned and the fine tuning of everyday operations within five days of our return from Elmendorf was complete.

"The UEI Team stayed for four days. The result of the teamwork during the UEI resulted in an overall squadron rating of EXCELLENT. These examples serve as a constant reminder to me of the incredible results that are possible when a unit works together as a team. This group of professionals banded together against time, the elements and the odds to prove that anything is possible when you do it as a team.

"While I am, extremely proud of the men and women of the 5072nd CSS I would also like to complement all the people of Elmendorf for their outstanding support and teamwork. It was their combined teamwork which enabled the airmen of Galena to overcome some incredible challenges and succeed as we have. Teamwork then is a very important ingredient in facing and overcoming tremendous challenges. The people at Galena know it works and so do the people at Elmendorf. We know how to do the Top Cover for North America Mission."

About the Publisher, Tactical 16

Tactical 16 Publishing is an unconventional publisher that understands the therapeutic value inherent in writing. We help veterans, first responders, and their families and friends to tell their stories using their words.

We are on a mission to capture the history of America's heroes: stories about sacrifices during chaos, humor amid tragedy, and victories learned from experiences not readily recreated—real stories from real people.

Tactical16 has published books in leadership, business, fiction, and children's genres. We produce all types of works, from self-help to memoirs that preserve unique stories not yet told.

You don't have to be a polished author to join our ranks. If you can write with passion and be unapologetic, we want to talk. Go to Tactical16.com to contact us and to learn more.

My hope is that the proceeds of this book will help the men and women who served this country when she called. I encourage other members of my community to tell their story.